2

COLLOQUIA

Shreeram S. Abhyankar
Mathematics Department
Purdue University
West Lafayette
IN 47907, USA

Pascal Auscher
Université de Paris-Sud
UMR du CNRS 8628
91405 Orsay Cedex, France

Fedor Bogomolov
Courant Institute of Mathematical Sciences, NYU
251 Mercer str.
New York, NY 10012, USA

John Coates
Emmanuel College
Cambridge CBZ 3AP, England

Klas Diederich
Bergische Universität
Gesanthochschule Wuppertal
Fachbereich 7 - Mathematik
Gaußstraße 20
42119 Wuppertal, Germany

Christopher D. Hacon
Department of Mathematics
University of Utah
Salt Lake City, UT 84112, USA

Roger Heath-Brown
Mathematical Institute
24–29, St. Giles'
Oxford OX1 3LB, UK

Luis Niremberg
Courant Institute
New York University
251 Mercer Street
New York, NY 10012

Claudio Procesi
Dipartimento di Matematica
"G. Castelnuovo"
P.le Aldo Moro, 2
00185 Roma, Italia

Vladimír Šverák
236 Vincent Hall
206 Church St. SE
University of Minnesota
Minneapolis, MN 55455, USA

Colloquium De Giorgi 2007 and 2008

Colloquium
De Giorgi
2007 and 2008

edited by
Umberto Zannier

EDIZIONI
DELLA
NORMALE

ISBN: 88-7642-344-4

Contents

Preface

Since 2001 the Scuola Normale Superiore di Pisa has organized the "Colloquio De Giorgi", a series of colloquium talks named after Ennio De Giorgi, the eminent analyst who was a Professor at the Scuola from 1959 until his death in 1996.

The Colloquio takes place once a month. It is addressed to a general mathematical audience, and especially meant to attract graduate students and advanced undergraduate students. The lectures are intended to be not too technical, in fields of wide interest. They must provide an overview of the general topic, possibly in a historical perspective, together with a description of more recent progress.

The idea of collecting the materials from these lectures and publishing them in annual volumes came out recently, as a recognition of their intrinsic mathematical interest, and also with the aim of preserving memory of these events.

For this purpose, the invited speakers are now asked to contribute with a written exposition of their talk, in the form of a short survey or extended abstract. After a first issue for the Colloquia of 2006, this is the second one, in a collection which we hope will continue on such yearly basis.

This volume contains a complete list of the talks held in the "Colloquio De Giorgi" in 2007, 2008 and also in the past years, and a table of contents of the first volume too.

Colloquia held in 2001

Paul Gauduchon
Weakly self-dual Kähler surfaces

Tristan Rivière
Topological singularities for maps between manifolds

Frédéric Hélein
Integrable systems in differential geometry and Hamiltonian stationary Lagrangian surfaces

Jean-Pierre Demailly
Numerical characterization of the Kähler cone of a compact Kähler manifold

Elias Stein
Discrete analogues in harmonic analysis

John N. Mather
Differentiability of the stable norm in dimension less than or equal to three and of the minimal average action in dimension less than or equal to two

Guy David
About global Mumford-Shah minimizers

Jacob Palis
A global view of dynamics

Alexander Nagel
Fundamental solutions for the Kohn-Laplacian

Alan Huckleberry
Incidence geometry and function theory

Colloquia held in 2002

Michael Cowling
Generalizzazioni di mappe conformi

Felix Otto
The geometry of dissipative evolution equations

Curtis McMullen
Dynamics on complex surfaces

Nicolai Krylov
Some old and new relations between partial differential equations, stochastic partial differential equations, and fine properties of the Wiener process

Tobias H. Colding
Disks that are double spiral staircases

Cédric Villani
When statistical mechanics meets regularity theory: qualitative properties of the Boltzmann equation with long-range interactions

Colloquia held in 2003

John Toland
Bernoulli free boundary problems - progress and open questions

Jean-Michel Morel
The axiomatic method in perception theory and image analysis

Jacques Faraut
Random matrices and infinite dimensional harmonic analysis

Albert Fathi
C^1 *subsolutions of Hamilton-Iacobi Equation*

Hakan Eliasson
Quasi-periodic Schrödinger operators - spectral theory and dynamics

Yakov Pesin
Is chaotic behavior typical among dynamical systems?

Don B. Zagier
Modular forms and their periods

David Elworthy
Functions of finite energy in finite and infinite dimensions

Colloquia held in 2004

Jean-Christophe Yoccoz
Hyperbolicity for products of 2×2 matrices

Giovanni Jona-Lasinio
Probabilità e meccanica statistica

John H. Hubbard
Thurston's theorem on combinatorics of rational functions and its generalization to exponentials

Marcelo Viana
Equilibrium states

Boris Rozovsky
Stochastic Navier - Stokes equations for turbulent flows

Marc Rosso *Braids and shuffles*

Michael Christ
The d-bar Neumann problem, magnetic Schrödinger operators, and the Aharonov-Böhm phenomenon

Colloquia held in 2005

Louis Nirenberg
One thing leads to another

Viviane Baladi
Dynamical zeta functions and anisotropic Sobolev and Hölder spaces

Giorgio Velo
Scattering non lineare

Gerd Faltings
Diophantine equations

Martin Nowak
Evolution of cooperation

Peter Swinnerton-Dyer
Counting rational points: Manin's conjecture

François Golse
The Navier-Stokes limit of the Boltzmann equation

Joseph J. Kohn
Existence and hypoellipticity with loss of derivatives

Dorian Goldfeld
On Gauss' class number problem

Colloquia held in 2006

Yuri Bilu
Diophantine equations with separated variables

Corrado De Concini
Algebre con tracce e rappresentazioni di gruppi quantici

Zeev Rudnick
Eigenvalue statistics and lattice points

Lucien Szpiro
Algebraic Dynamics

Simon Gindikin
Harmonic analysis on complex semisimple groups and symmetric spaces from point of view of complex analysis

David Masser
From 2 to polarizations on abelian varieties

Colloquia held in 2007

Klas Diederich
Real and complex analytic structures

Stanislav Smirnov
Towards conformal invariance of 2D lattice models

Roger Heath-Brown
Zeros of forms in many variables

Vladimir Sverak
PDE aspects of the Navier-Stokes equations

Christopher Hacon
The canonical ring is finitely generated

John Coates
Elliptic curves and Iwasava theory

Colloquia held in 2008

Claudio Procesi
Funzioni di partizione e box-spline

Pascal Auscher
Recent development on boundary value problems via Kato square root estimates

Hendrik W. Lenstra
Standard models for finite fields

Jean-Michel Bony
Generalized Fourier integral operators and evolution equations

Shreeram S. Abhyankar
The Jacobian conjecture

Fedor Bogomolov
Algebraic varieties over small fields

Louis Nirenberg
On the Dirichlet problem for some fully nonlinear second order elliptic equations

Contents of previous volume:

Colloquium De Giorgi 2006

Colloquium
De Giorgi

2007

Real and complex analyticity

Klas Diederich

1. Motivation and questions

Let X be a real analytic subset of some open set U in $\mathbb{C}^n$ or in a finite dimensional real analytic space. It has turned out, that it is important to study the following

Question. What are the relations of X to complex-analytic sets, for instance,

- What is the smallest complex-analytic set in $U \subset \mathbb{C}^n$ containing X (Consider the pair $\mathbb{R}^n \subset \mathbb{C}^n$)?
- Is X always the real part of some complex-analytic set Y in some $\mathbb{C}^N$? What is the smallest possible N?
- What are the largest complex-analytic sets through a given point $p \in X$ contained in X?
- Put

$$\mathcal{A} :=$$
$$\left\{ p \in X : \exists \, Y_p \subset X \text{ germ of compl.-anal. set with } p \in Y_p, \dim_p Y_p \geq 1 \right\}. \quad (1.1)$$

  What is the structure of $\mathcal{A}$? How large are the analytic sets in $\mathcal{A}$? Is $\mathcal{A}$ closed? Is it even real-analytic?

Most of the time, we do not even want to suppose that X is smooth, when asking these questions!

2. The special case of smooth X. The linear structure. Notions of complexifications

In this section it is supposed in general, that X is smooth, *i.e.*, a $\mathcal{C}^\omega$ real submanifold of U. For the time being the codimension of X can however be arbitrary. Let us denote by m the dimension of X.

In this case we can consider **tangent spaces** and **Levi forms**.

Definition 2.1. $p \in X$;

1. $T_p X$ *real tangent space* to X in p; it is a vector space $/\mathbb{R}$;
2. $T_p^h X := T_p X \cap J T_p X$ *the (real) holomorphic tangent space* (with J the standard complex structure on $\mathbb{C}^n$; it is a vector space $/\mathbb{R}$;
3. $T_p^{1,0} := \{X + iJX : X \in T_p X\}$ *the complex holomorphic tangent space* to X in p; it is a vector space $/\mathbb{C}$;
4. $T_p^{0,1} := \{X - iJX : X \in T_p X\}$ *complex antiholomorphic tangent space* to X in p; it is a vector space $/\mathbb{C}$.

Observation: One has

- $d := \dim_{\mathbb{R}} T_p X = \dim_{\mathbb{R}} X$.
- $k := \dim_{\mathbb{R}} T_p^h X$ always is even, $k = 2l$ (because $J^2 = -\mathrm{id}$, and, hence, $J(T_p^h X) = T_p^h X$.
- The real tangent space to X at p has the structure $T_p X = T_p^h X \oplus N$ with a real vector space N with $e := \dim_{\mathbb{R}} N = d - k$ and $T_p X \cap J N = \{0\}$.
- There is a real vector space E, $\dim_{\mathbb{R}} E = 2n - k - 2e$ and $JE = E$, such that
$$\mathbb{C}^n \cong \mathbb{R}^{2n} = T_p^h X \oplus N \oplus JN \oplus E$$
  N is called a totally real part of $T_p X$ (or *bad* part). X is totally real in a neighborhood of p, if $N = T_p X$, N is maximally totally real if it is totally real and if $E = \{0\}$.
- X is called generating (or generic) in a neighborhood of p if $E = \{0\}$, i.e., $2l + 2e = 2n$.
- More precisely, it can be said, that the smallest complex analytic subvariety $A \subset \mathbb{C}^n$ with $X \subset A$ must have complex dimension $\geq l + e$. So, a necessary condition for its existence is the inequality

$$l + e \leq n.$$

If its dimension is exactly $= l + e$ and it does not have unnecessary irreducible components, it might said to be generated by X or called **the (inner) complexification of X.** Trivially, if $l + e = n$, we have $A = \mathbb{C}^n$. An important question to be asked, however, is:

Question 2.2. Suppose that $l + e < n$. What is the inner complexification of X in this case?

In the last item the word "inner" was added to the definition of a complexification, since there also always is another complexification, usually just called complexification, but which will be called "exterior complexification" in our terminology. It can be characterized in the following way:

Definition 2.3. Let X just be a real-analytic set in $\mathbb{C}^n \cong \mathbb{R}^{2n}$. The complex-analytic set $\tilde{X} \subset \mathbb{C}^{2n}$ is called an exterior complexification of X, if X considered as a subset of the space $\mathbb{R}^{2n} = \Re\mathbb{C}^{2n}$ satisfies

(a) X is a closed subset of $\tilde{X}$ and $X = \tilde{X} \cap \mathbb{R}^{2n}$,
(b) $\mathcal{O}(\tilde{X}_x) = \mathcal{O}(X_x) \otimes_{\mathbb{R}} \mathbb{C} \; \forall x \in X$.

Remark 2.4. Because of (a), X is uniquely determined by any exterior complexification $\tilde{X}$ of X. In this sense, X can be called the totally real part of $\tilde{X}$. This is not the case for inner complexifications as the trivial situation $l + e = n$ shows. In this case $\mathbb{C}^n$ is always the universal inner complexification which does no longer depend on X.

One has

Proposition 2.5. *Let X be a real (coherent) analytic set in an open set $\Omega \subset \mathbb{R}^{2n}$. Then:*

1. *Existence: There exists a complexification $\tilde{X}$ of X,*
2. *Uniqueness: If $\tilde{X}_1$ and $\tilde{X}_2$ are exterior complexifications of X, then $\tilde{X}_1 \cap \tilde{X}_2$ is an exterior complexification of X.*
3. *X has in $\tilde{X}$ a Stein neighborhood basis.*

Remark 2.6. "Coherent" means that the sheaf of real-analytic functions on X is coherent.

3. Complex-analytic sets in real-analytic ones and their relevance for the $\overline{\partial}$-problem

Next, we come to the inverse situation, namely, we consider complex-analytic sets inside real-analytic ones. One has the following fact (the proof is easy)

Proposition 3.1. *Let $A \subset X$ be a complex-analytic submanifold and $p \in A$ (important: we do not suppose, that $\dim A = \dim T_p^{1,0} X$). Then the vector-valued Levi-form $\mathcal{L}_X(L, \overline{L'}; p) = 0 \; \forall L \in T_p^{1,0} A$ and $\forall L' \in T_p^{1,0} X$. In other words, let*

$$\mathcal{N}_p X := \{ L \in T_p^{1,0} X : \mathcal{L}_X(L, \overline{L'}; p) = 0 \quad \forall L' \in T_p^{1,0} X \}$$

be the null space of the Levi form of X at p. Then one has

$$T_p^{1,0} A \subset \mathcal{N}_p X. \tag{3.1}$$

This says, that complex-analytic sets A inside X enforce strong Levi degeneracies of X along A. A question important for deep analytic applications (see J. J. Kohn [14]) now is:

Question 3.2. In which sense does the converse hold true? More precisely: Which kind of Levi degeneracy of X implies the existence of complex analytic submanifolds in X of positive dimension without necessarily enforcing a complex-analytic foliation on X?

J. J. Kohn considered in [14] the following situation and notions:

Definition 3.3. Let $V \subset \mathbb{C}^n$ be a real-analytic subvariety (possibly with singularities) and let $p \in V$ be an arbitrary point.

1. By $\mathcal{I}_p(V)$ we denote the ideal of germs at p of $\mathcal{C}^\omega$ functions vanishing on V.
2. The complex vector space

$$T_p^{1,0} V := \left\{ T = \sum_{i=1}^n t_i \frac{\partial}{\partial z_i} \Big| (Tf)(z) = 0 \, \forall z \in \mathcal{I}_p(V) \right\}$$

   is called the (complex) holomorphic Zariski tangent space of V at p.

From now on in this section we denote by S **a pseudoconvex $\mathcal{C}^\omega$-smooth hypersurface in an open set $\Omega \subset \mathbb{C}^n$ and for $z \in S$ by $L_S(z; \cdot)$ the Levi-form of S at z with respect to any defining function.** Let $\mathcal{N}_z \subset T_z^{1,0} S$ be the null-space of $L_S(z; \cdot)$ at z. Then J. J. Kohn defined in [14]:

Definition 3.4.

1. Let $V \subset S$ be a closed real-analytic subvariety and $z \in V$. The number

   $$\mathrm{hol\,dim}_{z,S} V = \sup_{U=U(z)} \inf_{w \in U \cap V} \dim_{\mathbb{C}}(T_w^{1,0} V \cap \mathcal{N}_w)$$

   where U runs over all open neighborhoods of z, is called the **holomorphic dimension** of V at z in S.
2. The *holomorphic dimension* of V in S is

   $$\mathrm{hol\,dim}_S V = \min_{z \in V} \mathrm{hol\,dim}_{z,S} V .$$

Remark 3.5. If $\mathrm{hol\,dim}_{z_0} V = q$, every neighborhood of $z_0 \in V$ obviously contains a regular point $z \in V$ such that $\dim_{\mathbb{C}}(T_w^{1,0}) \cap \mathcal{N}_w = q$ for all $w \in V$ near z.

Proposition 3.1, therefore, immediately gives

Proposition 3.6. *Let $A_z \subset S$ be a germ of a complex-analytic set. Then one has*

$$\dim_{\mathbb{C}} A_z = \mathrm{hol\,dim}_S A_z . \tag{3.2}$$

The notion of holomorphic dimension is relevant because J. J. Kohn showed in [14] that it naturally appears when the existence of subelliptic estimates for the $\bar\partial$-operator is investigated near pseudoconvex real-analytic boundaries using the **theory of subelliptic multipliers**. Kohn showed:

Theorem 3.7 (J. J. Kohn). *Suppose $\Omega \subset \subset \mathbb{C}^n$ is pseudoconvex with smooth boundary, $z_0 \in \partial\Omega$, and there is a neighborhood U of z_0 such that $\partial\Omega \cap U$ is C^ω and satisfies the following condition:*

$$\nexists \quad V \subset \partial\Omega \cap U \text{ germ of } C^\omega\text{-subvariety with } \mathrm{hol\,dim}\, V \geq q. \qquad (\mathbf{I_q})$$

Then subelliptic estimates for the $\bar\partial$-Neumann problem for (p,q)-forms on Ω hold near all points of $\partial\Omega \cap U$.

A subelliptic estimate of order ε for the $\bar\partial$-Neumann problem holds on a pc. domain $\Omega \Subset \mathbb{C}^n$ with C^∞-boundary, if the following estimate holds:

$$\|f\|_{\varepsilon,\Omega}^2 \leq C \left(\left\| \bar\partial f \right\|_\Omega^2 + \left\| \bar\partial^* f \right\|_\Omega^2 \right) \qquad (3.3)$$

$\forall f \in \mathrm{Dom}(\bar\partial) \cap \mathrm{Dom}(\bar\partial^*)$.
A consequence of the existence of a subelliptic estimate of order ε at a point $z \in \partial\Omega$ (meaning on the intersection of Ω with a suitable neighborhood of z) is that the $\bar\partial$-problem can be solved near z with a regularity gain of ε in the degree of the Sobolev norm.

Because of the result of Kohn it is particularly important to know that the converse of Proposition 3.6 also holds. More precisely, one has:

Theorem 3.8 (Die-Fornæss [11]). *Let S be a pseudoconvex C^ω-smooth hypersurface in an open set $W \subset \mathbb{C}^n$. Suppose $V \subset S$ is a C^ω subvariety (not necessarily closed) with $\mathrm{hol\,dim}\, V = q$. Furthermore, let $z_0 \in V$ and $U = U(z_0) \subset W$ be an arbitrary open neighborhood. Then there is a complex-analytic submanifold $A \subset U \cap S$ with $\dim_\mathbb{C} A = q$. The manifold A always can be chosen such that $A \cap V \neq \emptyset$.*

Remark 3.9. It cannot be expected that $A \subset V$ as the following example shows:
Put $S := \{\mathrm{Im}\, z_n = 0\}$ and $V := \{\|z\| = 1\} \cap S$. Since S is totally Levi flat, one has

$$\mathrm{hol\,dim}\, V = n - 2.$$

Nevertheless, V does not contain any germ of a complex-analytic set A of dimension $n-2$, since it is contained in a strictly pseudoconvex sphere in $\mathbb{C}^n$. (Notice, however, that as A any of the complex leaves of S can be taken.)

The remark says, in particular, that Theorem 3.8 cannot be proved by integration of vector fields on and tangent to V. Instead, some deeper analysis is necessary for the proof, in which, in fact, the complex manifold is at first constructed in such a way that it passes through a point $z \in S$ arbitrarily close to z_0 and has infinite order of contact with S. Because of the inequality of Łojaziewicz it, therefore, has to be contained in S.

4. Real-analytic varieties and their complex analytic subvarieties: general local case

We come back to the case of an arbitrary closed $\mathcal{C}^\omega$-subvariety X of an open set $W \subset \mathbb{C}^n$.

Let $z_0 \in X$. We make no assumption on the dimension of X. It also might have singularities. There is a radius $\varrho > 0$ and on the polycylinder $\Delta_\varrho(z_0)$ of radius ϱ around z_0 a $\mathcal{C}^\omega$-function $r : \Delta_\varrho(z_0) \to \mathbb{R}$, such that

$$X \cap \Delta_\varrho = \{z \in \Delta_\varrho : r(z) = 0\}. \tag{4.1}$$

In [11] the following crucial Lemma was shown:

Lemma 4.1. *Let X, z_0, ϱ be as above. Suppose, there is a germ of a complex-analytic set A_{z_0} at z_0 with $A_{z_0} \subset X$ and $\dim A_{z_0} > 0$. Then there is a closed complex-analytic subset $\hat{A} \subset X \cap \Delta_\varrho(z_0)$ such that*

$$\dim A_{z_0} \leq \dim_{z_0} \hat{A}. \tag{4.2}$$

Remark 4.2. The point of the Lemma is, that $\hat{A}$ is closed in $\Delta_\varrho(z_0)$), whereas the given germ A_{z_0} might have a reprensentative only in a very small open set. In this sense the Lemma guarantees the existence of a complex-analytic set extending far enough. Because the germ A_{z_0} itself might, indeed, not extend very far, it can, in general, not be expected, that $A_{z_0} \subset \hat{A}$ and both have the same dimension. The dimension of $\hat{A}$ might, indeed, have to be larger than the dimension of A_{z_0}.

As mentioned the Lemma has been proved already by Diederich/Fornaess in [11]. Here, we want to point out the particular role of Segre varieties for the existence of $\hat{A}$. We recall

Definition 4.3. Let X, z_0, r, ϱ be as above. Then, after suitably shrinking ϱ the real-analytic function $r(z)$ has a natural complexification which we denote by $r(z, w)$. It is defined on the direct product $\Delta_\varrho(z_0)(z) \times \Delta_\varrho(z_0)(w)$ and is holomorphic in z and antiholomorphic in w. Let now

$w \in X \cap \Delta_\varrho(z_0)$ be an arbitrary fixed point. Then we call the closed complex-analytic subset of $\Delta_\varrho(z_0)$

$$\hat{S}_w := \{z \in \Delta_\varrho(z_0) : r(z, w) = 0\}$$

a **Segre hypersurface associated to** X **at** w.

Notice, that $\hat{S}_w(z_0)$ might have singularities (One even might have $\hat{S}_w(z_0) = \mathbb{C}^n$). Furthermore, it is important to observe, that the function r is not uniquely determined by X and z_0 up to a non-vanishing real-analytic function as a factor (like it is in the case of a smooth hypersurface). Hence, the Segre hypersurface is not canonically associated to the pair (X, z_0). Another disadvantage of the $\hat{S}_w$ is, that its codimension in $\mathbb{C}^n$ always is ≤ 1 even if X itself has higher codimension. Nevertheless, the rather crude notion of Segre varieties for singular real-analytic sets given by the definition above turns out to be very useful. In particular, suitable intersections of Segre hypersurfaces can be used for the proof of Lemma 4.1. We will indicate next, how this can be done. The most crucial Lemma for the proof of Lemma 4.1 is the following fact:

Lemma 4.4 (Principal lemma). *Assume that* $0 \in X$ *and let* r *be a defining function of* X *in a neighborhood of* 0 *with convergent power series around* 0 *on the polydisk* $\Delta^n(0, \varrho)$ *at* 0 *of radius* ϱ. *If* $0 \in \gamma \subset X$ *is a piece of a non-constant holomorphic disc with holomorphic parametrization* $z(t)$. *Then*

$$\gamma \subset S_0(X) := \{z \in \Delta^n(0, r) \mid \hat{r}(z, 0) = 0\} \tag{4.3}$$

where $\hat{r}$ *is the complexification of* r.

Proof. We reorder the Taylor series expansion of the function $\hat{r}(z(t), z(t))$ around the point $(z(t), 0)$ a little bit and get

$$\underbrace{\hat{r}(z(t), \overline{z(t)})}_{\equiv 0,\ \text{since}\ \gamma \subset X} - \underbrace{\hat{r}(z(t), 0)}_{\text{hol. in } t} = \underbrace{\sum_{|\alpha|=1}^{\infty} \frac{1}{\alpha!} \partial_{\bar{w}}^\alpha \hat{r}(z(t), 0)\overline{z(t)}^\alpha}_{\text{each term has factor of } \bar{t}} \tag{4.4}$$

Since the left-hand side is holomorphic in t and all terms on the right hand side contain $\bar{t}$, there cannot be any non-vanishing term. Hence we get $\hat{r}(z(t), 0) \equiv 0$. □

As a consequence of this lemma one obtains as we will see a first structure theorem on the set $\mathcal{A}$ as introduced in (1.1) the following:

Theorem 4.5. *The subset* $A \subset X$ *as introduced in* (1.1) *carries naturally the structure of a lamination (with singularities) by complex-analytic manifolds.*

Another very important application of Lemma 4.1 whose now simple proof has first been given in [11] is

Theorem 4.6. *If the given real-analytic set* X *is compact, then* $A = \emptyset$.

Theorems 3.8, 3.7 and 4.6 together immediately give the

Corollary 4.7. *Let* $\Omega \subset \mathbb{C}^n$ *be a bounded pseudoconvex domain with smooth* C^ω *boundary. Then for all* $1 \leq q \leq n$ *subelliptic estimates hold on* Ω *for the* $\bar{\partial}$*-Neumann problem on* $(0, q)$*-forms on* Ω.

Remark 4.8.

i) The use of theorems 3.7 and 3.8 in this conclusion can be replaced by using the work of D. Catlin (see [5] and [6]). However, also in this procedure it has to be known, that the given domain Ω automatically is of finite type, which - in the real-analytic case - is equivalent to knowing $A = \emptyset$. In other words, theorem 4.6 and, hence, also Lemma 4.1 still enter into the conclusion. It is nowadays usual to follow this alternative.

ii) The path of argumentation in 4.8) is usually preferred since it also covers in its correct interpretation the case of bounded C^∞-smooth domains (of finite type). However, there are 2 disadvantages attached to it:

   a) Catlin's proof in [6] is highly complicated. It would be extremely nice to obtain a simplification and, maybe, also clarification of it.

   b) An essential desire is to obtain an explicit estimate for the best (largest) possible ε for which subelliptic estimates for the $\bar{\partial}$-Neumann problem hold on Ω in terms of geometrical data (like the types in the sense of d'Angelo (see [7]) or multitypes in the sense of Catlin (see [4])). This does not seem to be possible using the method of D. Catlin (the estimates are terribly weak). There is hope, that the theory of subelliptic multipliers as developed by J. J. Kohn in [14] can give precise estimates for the degree of subellipticity. Besides J. J. Kohn and some of his students, Y. T. Siu is following this project together with a whole group of others(See the recent preprint of Y.T. Siu [16]). An advantage of this approach also might be, that it might give at the same time important quantitative information for the multiplier ideal theory as used in algebraic geometry by A. Nadel in [15] and then by Y. T. Siu, J. P.

Demailly, J. Kollár (see [10] for a very good survey). In particular, hope is, that a quantitative version of the Hironaka theorem might come out of this.

5. The closedness of $\mathcal{A}$

Let S be a $\mathcal{C}^\infty$-smooth pseudoconvex hypersurface (not necessarily compact) and let $p \in S$ be a point. Suppose, there is a germ A_p at p of a positive-dimensional complex-analytic set, such that $A_p \subset S$. It was already shown by Diederich/Pflug in [12] and, independently by D. Catlin in [3], that in this case subelliptic estimates for the $\bar{\partial}$-Neumann-problem on the pseudoconvex side of S do not hold. This fact together with Corollary 4.7 allows to conclude:

Proposition 5.1. *Let S be a non-compact $\mathcal{C}^\omega$-smooth pseudoconvex hypersurface. Then the set $\mathcal{A}(S)$ is closed.*

Namely, according to the above-mentioned facts the set $S \setminus \mathcal{A}$ consists exactly of those points on S at which subelliptic estimates hold. This set, however, is trivially open, such that $\mathcal{A}$ is necessarily closed.

Remark 5.2.

a) The argument given above only seems to be short. In reality, a whole lot of complicated matters go into it. Therefore, a simplification is desirable.

b) The hypotheses of the proposition are not at all natural. Why should pseudoconvexity play a role? Why smoothness? Why the fact, that $\operatorname{codim}_{\mathbb{R}} S = 1$?

c) It should be mentioned that the questions in b) have been asked by many people, also by John P. d'Angelo. In fact, he states in [8] and in [9] that $\mathcal{A}$ is closed. However, he does not state the hypotheses very clearly and there is no explicit proof. His indications for a possible argument, in any case, would give a very non-constructive, indirect proof (what this means, will become clear from the sketch of arguments below .)

In 2006, however, the author showed together with E. Mazzilli, Lille, the following

Theorem 5.3. *Let $X \subset U \subset \mathbb{C}^n$ be a closed real-analytic set in the open set U in $\mathbb{C}^n$. Then the set $\mathcal{A}(X)$ is closed as a subset of U.*

Remark 5.4. The result will be published in Diederich/Mazzilli: "Real and Complex analytic sets. The relevance of Segre varieties". (To appear in Annali della Scuola Normale Superiore di Pisa. Classe di Scienze 2008.)

In trying to prove Theorem 5.3 one might try to do the following:

Let p_0 be an accumulation point of $\mathcal{A}(X)$. We want to show, that $p_0 \in \mathcal{A}$, i.e., we want to find a positive-dimensional complex-analytic set $A_{p_0} \subset X$ passing through p_0. (Notice that there will be such a set which is even closed in $V = V(p_0)$ if there is any (see Lemma 4.1)). We can pick a sequence of points $p_k \in \mathcal{A}(X)$ which converges to p_0 and, due to the Principal Lemma (see Lemma 4.1) through any p_k passes a positive-dimensional complex analytic closed subset $A_k(X)$ of the fixed neighborhood $V = V(p_0)$. Since, hence, all the A_k have in some sense the same size, they cannot shrink down to p_0. Hence one might conjecture, that the limit set

$$\hat{A}_0 := \lim_{k \to \infty} A_k$$

must contain the required set A_0. However, the famous counterexample of Stolzenberg shows, that this cannot be guaranteed without further information. The main obstacle lies in the fact, that one might have

$$\limsup_{k \to \infty} \operatorname{vol} A_k = \infty .$$

E. Bishop, however, showed in [2] the following:

Theorem 5.5 (E. Bishop). *Let $A_\nu \subset U \subset \mathbb{C}^n$ be a sequence of purely p-dimensional analytic subsets converging to some set $A \subset U$ and such that for any compact subset $K \subset U$ there is a constant $M_K \geq 0$ with*

$$\operatorname{vol}_{2p}(A_\nu \cap K) \leq M_K \quad \forall \nu .$$

Then also A is a purely p-dimensional complex-analytic subset of U.

This implies immediately

Corollary 5.6. *Let $A_\nu \subset U$ be a sequence of purely p-dimensional complex-analytic subsets with locally uniformly bounded $2p$-dimensional Hausdorff measures:*

$$\operatorname{vol}_{2p}(A_\nu \cap K) \leq M_K \quad \forall \nu$$

for a suitable constant M_K associated to any compact subset $K \subset U$. Then we can extract a subsequence from (A_ν) converging in U to a purely p-dimensional complex-analytic subset $A \subset U$ or to $\emptyset$.

The hypothesis of the Bishop-Theorem and its Corollary is, in general, difficult to verify. However, in trying to prove Theorem 5.3 this difficulty can be overcome by showing at first, that there is, in fact, a close relation between Segre varieties $S_p(X)$ of X and a suitable choice of the A_k. The advantage of this is, that the S_{p_k} depend antiholomorphically on the parameter p_k. Such a situation has been studied by Die/Pinchuk in [13]. They showed, that analytic dependence on the parameter of a sequence of complex-analytic sets allows to deduce the local boundedness of the volume. In order to state their result precisely we give at first the following

Definition 5.7. We say, that a sequence of subsets $E_j \subset U$ converges to a set $E \subset U$ if E consists of all limit points of convergent sequences x_{v_j} with $x_{v_j} \in E_{v_j}$ and if there is for any compact subsets $K \subset E$ and any $\varepsilon > 0$ an index $v(\varepsilon, K)$ such that K belongs to the ε-neighborhood of E_v in U for all $v > v(\varepsilon, K)$.

The main result of [13] then is

Theorem 5.8. *Let* $U \subset \mathbb{C}^n$, $V \subset \mathbb{C}^m$ *be open sets and* $g_j(z, w) \in \mathcal{O}(U \times V)$ *for* $j = 1, \ldots, k$ *with a positive integer* $k \leq n$. *For* $w \in V$ *put*

$$A_w := \{ z \in U : g_j(z, w) = 0 \text{ for } j = 1, \ldots, k \}.$$

Let $E := \{ w \in V . \dim_{\mathbb{C}} A_w > n - k \}$. *Then for any* $\tilde{U} \Subset U$, $\tilde{V} \Subset V$ *there exists a constant* $c = c(\tilde{U}, \tilde{V}) > 0$ *such that*

$$\text{vol}_{2(n-k)} \left(A_w \cap \tilde{U} \right) < c$$

for all $w \in \tilde{V} \setminus E$. *In particular, we can extract from any sequence* (A_{w_v}), $w_v \in \tilde{V} \setminus E$, *a subsequence converging in* U *to an analytic subset* A *of pure dimension* $n - k$. *(Notice, that the sequence* (w_v) *might converge to a point in* E.)

Remark 5.9. D. Barlet pointed out to the author during the conference on Complex Analysis in Bucharest in June 2008, that this result is already contained in Theorem 3 of his paper [1].

Putting this simple idea to work, is technically rather complicated. We explain here the main steps.

5.1. Intersections of Segre varieties

At first it can be shown by a repeated application of the Principal Lemma (4.4), that the following holds true:

Lemma 5.10. *Let as above* $a_N \in A$ *be a convergent sequence,* $\lim a_N = a_0$. *Then there are for any index* N *a number* l_N *and points* z_j^N, $j =$

$1, \ldots, l_N$ in X, such that one has

$$A_N := \bigcap_1^{l_N} S_{z_j^N} \subset X.$$ (5.1)

It also may be assumed that the sets A_N are all pure-dimensional of dimension p, independent of N.

Remark 5.11. Notice, that the complex-analytic sets A_N are as finite intersections of Segre varieties of X closed subsets of a fixed neighborhood $U(p_0)$.

A next important step is to deduce from this

Lemma 5.12. *There exists an integer l such that for each N a suitable choice of l points $(w_1^N, \ldots, w_l^N) \subset (z_1, \ldots, z_{l_N}^N)$ can be made such that one has for*

$$C_N := \bigcap_1^{l} S_{w_j}^N$$ (5.2)

$$\dim C_N = \dim A_N.$$ (5.3)

Furthermore, one has, of course,

$$A_N \subset C_N.$$ (5.4)

Remark 5.13. The complete proof of this statement involves already a first application of Theorem 5.8.

5.2. Uniform boundedness of volumes

We want to show, that the volume of the A_N is uniformly bounded in N. Then it follows from Bishop's theorem that a suitable subsequence of (A_N) converges to a p-dimensional complex-analytic set $A \subset X$ whith $0 \in A$. Hence we get that $0 \in \mathcal{A}$.

In order to show this it suffices because of (5.4) and (5.3), of course, to show that the $2p$-dimensional volumes of the sets C_N are uniformly bounded.

The trick how to reduce this to an application of Theorem 5.8 consists in considering for a suitable small radius $r > 0$ and a neighborhood $\tilde{U}$ of a_0 the mapping

$$G : \tilde{U} \times (B_{r+\varepsilon})^l \times \mathbb{C}^{l(n-p)} \to \mathbb{C}^{n-p}$$ (5.5)

$$(z, w_1, \ldots, w_l, \lambda^1, \ldots, \lambda^{n-p}) \to \left(\sum_{j=1}^{l} \lambda_j^1 \hat{\varrho}(z, w_j), \ldots, \sum_{j=1}^{l} \lambda_j^{n-p} \hat{\varrho}(z, w_j) \right)$$ (5.6)

which is analytic in all its variables. (It satisfies all requirements of Theorem 5.8). This finishes the proof of Proposition 5.1.

6. Open questions

As mentioned already in Section 1 it would be desirable to know more about the structure of the set $\mathcal{A}$ for an arbitrary real-analytic subset $X \subset U \subset \mathbb{C}^n$ of an open set U of $\mathbb{C}^n$. The best one might hope for is that $\mathcal{A}$ necessarily is real-analytic itself. The proof of this conjecture might be quite difficult in the general case considered here. Of course, the conjecture also could be considered first in the case, when X is a pseudoconvex $\mathcal{C}^\omega$ smooth hypersurface.

Another open question also has to be considered in this context. Namely, let us say, that X is of infinite type at some point $p \in X$, if for any given positive integer m there is a closed complex-analytic curve $A \ni p$ of a suitable open neighborhood U of p (which might depend on m), such that the order of contact between A and X at p is larger than the given m.

Notice at first that X is, of course, of infinite type at all points of a complex-analytic curve $A \subset X$. So the question is, whether the converse also holds true, namely

Question. Let X be of infinite type at $p \in X$. Does p necessarily lie on a non-constant complex curve $A \subset X$?

This question is intimately related to the following:

Question. If X has 1-type $\geq m$ at a point $p \in X$. Is there a suitable Segre variety $S_q(X)$ passing through p and containing a non-constant complex curve $A \subset X$ through p with order of contact with X at $p \geq m$.

In order to answer this last question a new notion of Segre varieties for singular real-analytic sets might have to be found.

References

[1] D. BARLET, *Majoration du volume des fibres génériques et forme géométrique du théorème d'aplatissement*, In: "Séminaire Pierre Lelong/Henri Skoda" (Analyse), 1978/79, Lecture Notes in Math., Vol. 822, Springer, Berlin-New York.

[2] E. BISHOP, *Condition for the analyticity of certain sets*, Michigan Math. J. **11** (1964), 289–304.

[3] D. CATLIN, *Necessary conditions for subellipticity and hypoellipticity for the $\bar{\partial}$-Neumann problem on pseudoconvex domains*, Recent developments in several complex variables, Proc. Conf., Princeton Univ. 1979 (Fornæss, J. E., ed.), Ann. Math. Stud. 100, 93–100.

[4] D. CATLIN, *Necessary conditions for subellipticity of the $\bar{\partial}$-Neumann problem*, Ann. Math. **117** (1983), 147–171.

[5] D. CATLIN, *Boundary invariants of pseudoconvex domains*, Ann. Math. **120** (1984), 529–586.

[6] D. CATLIN, *Subelliptic estimates for the $\bar{\partial}$-Neumann problem on pseudoconvex domains*, Ann. Math. **126** (1987), 131–191.

[7] J. D'ANGELO, *Finite type conditions for real hypersurfaces*, J. Diff. Geo. **14** (1979), 59–66.

[8] J. D'ANGELO, *Real hypersurfaces, orders of contact, and applications*, Ann. Math. **115** (1982), 615–637.

[9] J. P. D'ANGELO, Stud. Adv. Math., CRC Press, Boca Raton, FL, 1993.

[10] J. P. DEMAILLY, *Multiplier ideal sheaves and analytic methods in algebraic geometry*, School on vanishing theorems and effective results in algebraic geometry, Lecture notes of the school held in Trieste, Italy, April 25-May 12, 2000. ICTP Lecture Notes 6. (Trieste) (Demailly, J. P. et al., ed.), The Abdul Salam International Center for Theoretical Physics, 1–148.

[11] K. DIEDERICH and J. E. FORNÆSS, *Pseudoconvex domains with real analytic boundary*, Ann. Math. **107** (1978), 371–384.

[12] K. DIEDERICH and P. PFLUG, *Necessary conditions for hypoellipticity of the $\bar{\partial}$-problem*, In: "Recent developments in several complex variables" (Princeton, N.J.), J. E. Fornæss, (ed.), Annals of Mathematics Studies, Vol. 100, Princeton University Press, 151–154.

[13] K. DIEDERICH and S. PINCHUK, *Uniform volume estimates for holomorphic families of analytic sets*, In: "Proceedings of the Steklov Institute of Mathematics", Vol. 235 (2001) (Moscow), 52–56.

[14] J. KOHN, *Subellipticity of the $\bar{\partial}$-Neumann problem on pseudoconvex domains: Sufficient conditions*, Acta. Math. **142** (1979), 79–122.

[15] A. M. NADEL, *Multiplier ideal sheaves and Kähler-Einstein metrics of positive scalar curvature*, Ann. Math. **132** (1990), 549–596.

[16] Y. T. SIU, *Effective termination of Kohn's algorithm for subelliptic multipliers*, Preprint 2008.

Zeros of forms in many variables

Roger Heath-Brown

Let $F(x_1, \ldots, x_n) \in \mathbb{Z}[x_1, \ldots, x_n]$ be a form of degree d. A fundamental question in the theory of Diophantine equations is to determine when

$$F(x_1, \ldots, x_n) = 0$$

has a non-trivial solution in $\mathbb{Z}$. When $d = 2$ we have a complete answer to this question. Meyer showed in 1884 that if $d = 2$ and F is indefinite, then there is always a solution when $n \geq 5$. More generally we have the following result.

Theorem 1 (Hasse-Minkowski). *If $F(x_1, \ldots, x_n) \in \mathbb{Z}[x_1, \ldots, x_n]$ is an indefinite quadratic form, then there is a solution $F(x_1, \ldots, x_n) = 0$ with*

$$(x_1, \ldots, x_n) \in \mathbb{Z}^n - \{0\}$$

if and only if every congruence

$$F(x_1, \ldots, x_n) \equiv 0 \pmod{p^e} \tag{1}$$

with p prime and $e \in \mathbb{N}$ is solvable with $p \nmid$ h.c.f.$(x_1, \ldots, x_n)$.

The "only if" part of the the theorem is quite easy to see, but the reverse implication is harder.

It is natural to ask whether there are analogues of these results for higher degrees. When d is odd the equation automatically has real solutions, so that an indefiniteness condition is unnecessary. However it transpires that there can be congruence conditions for every prime p and every degree d. We illustrate this via the case $p = 2, d = 3$, starting from the observation that

$$(1+x_1)(1+x_2)(1+x_3) \equiv 1 \pmod 2 \implies x_1 \equiv x_2 \equiv x_3 \equiv 0 \pmod 2.$$

On expanding the left-hand side we find that if

$$x_1 + x_2 + x_3 + x_1 x_2 + x_1 x_3 + x_2 x_3 + x_1 x_2 x_3 \equiv 0 \pmod 2$$

then

$$x_1 \equiv x_2 \equiv x_3 \equiv 0 \quad (\text{mod } 2).$$

Since $x_i \equiv x_i^2 \equiv x_i^3$ (mod 2) for any integer x_i we deduce that

$$N(x_1, x_2, x_3) \equiv 0 \quad (\text{mod } 2) \quad \Rightarrow \quad x_1 \equiv x_2 \equiv x_3 \equiv 0 \quad (\text{mod } 2)$$

with

$$N(x_1, x_2, x_3) = x_1^3 + x_2^3 + x_3^3 + x_1^2 x_2 + x_1^2 x_3 + x_2^2 x_3 + x_1 x_2 x_3.$$

We now take

$$F(x_1, \ldots, x_9) = N(x_1, x_2, x_3) + 2N(x_4, x_5, x_6) + 4N(x_7, x_8, x_9).$$

Then $F(x_1, \ldots, x_9) = 0$ implies $2|N(x_1, x_2, x_3)$, whence $2|x_1, x_2, x_3$. It follows that $8|N(x_1, x_2, x_3)$. Thus $F(x_1, \ldots, x_9) = 0$ implies $4|2N(x_4, x_5, x_6)$ and we can repeat the argument to show that $2|x_4, x_5, x_6$. A further repetition of the argument shows that $2|x_7, x_8, x_9$. Thus the congruence (1) for $p=2$ and $e=3$ has no solution with $2 \nmid \text{h.c.f.}(x_1, \ldots, x_9)$.

In general there are versions of this construction for any prime and any degree d, producing a form $F(x_1, \ldots, x_n)$ with $n = d^2$ such that

$$F(x_1, \ldots, x_n) \equiv 0 \quad (\text{mod } p^d) \quad \Rightarrow \quad p|\text{h.c.f.}(x_1, \ldots, x_n).$$

Artin conjectured, in effect, that this is best possible.

Conjecture 1 (Artin). Let $F(x_1, \ldots, x_n) \in \mathbb{Z}[x_1, \ldots, x_n]$ be a form of degree d. If $n > d^2$ then for any prime p and any exponent $e \geq 0$ the congruence

$$F(x_1, \ldots, x_n) \equiv 0 \quad (\text{mod } p^e)$$

has an integral solution with $p \nmid \text{h.c.f.}(x_1, \ldots, x_n)$.

A better framework for such congruence conditions is provided by working in p-adic fields. For any prime p we define a metric on $\mathbb{Q}$ by setting

$$d_p(x, y) = \begin{cases} 0, & x = y, \\ p^{-v}, & x \neq y, \end{cases}$$

where $v \in \mathbb{Z}$ is the unique integer such that $x - y = p^v \frac{a}{b}$ with $a, b \in \mathbb{Z}$ and $p \nmid ab$. It is an easy exercise to verify that d_p is a metric. The p-adic field $\mathbb{Q}_p$ is the completion of $\mathbb{Q}$ under this metric, and the p-adic integers are the corresponding completion of $\mathbb{Z}$. We then have the following key fact.

Given a form $F(x_1, \ldots, x_n) \in \mathbb{Z}[x_1, \ldots, x_n]$ and a prime p, the condition (1) has a solution for every e, with $p \nmid \text{h.c.f.}(x_1, \ldots, x_n)$, if and only if if the equation $F(x_1, \ldots, x_n) = 0$ has a non-trivial solution in $\mathbb{Q}_p^n$ (or, equivalently, in $\mathbb{Z}_p^n$). Thus Artin's Conjecture may be re-formulated as stating that $F(x_1, \ldots, x_n)$ has a non-trivial p-adic zero as long as $n > d^2$.

We can now split our fundamental problem (on the solvability of the equation $F(x_1, \ldots, x_n) = 0$ over $\mathbb{Z}$) into two parts. Firstly we have a "p-adic problem", asking for which n the equation $F(x_1, \ldots, x_n) = 0$ always has non-trivial p-adic solutions. For example when $d = 2$ it suffices to have $n \geq 5$, and in general Artin's conjecture would say that $n \geq d^2 + 1$ suffices. Secondly we have the "local-to-global problem". This asks for which n the necessary p-adic condition (together with a suitable indefiniteness condition, when d is even) implies the existence of non-trivial solutions of $F(x_1, \ldots, x_n) = 0$ over $\mathbb{Z}$.

Thus the Hasse-Minkowski theorem gives a positive answer to the local-to-global problem for quadratic forms, for every $n \geq 1$. Naturally we would like to extend this to higher degrees, but there are well-known counter-examples

$$3x_1^3 + 4x_2^3 + 5x_3^3 = 0$$

and

$$5x_1^3 + 9x_2^3 + 10x_3^3 + 12x_4^3 = 0$$

in which the p-adic conditions are satisfied but there is only the trivial solution in $\mathbb{Z}$. In general we may hope for a modified version of the local-to-global principle, in which the p-adic condition is replaced by the stronger, but more subtle, Brauer-Manin condition. However we shall not go further into this.

The p-adic and local-to-global problems are solved for sufficiently large n by the following two results (at least for odd degrees, in the local-to-global case).

Theorem 2 (Brauer, 1945 [4]). *For every degree d there is an integer $n(d)$ such that for each prime p, every form $F(x_1, \ldots, x_n) \in \mathbb{Z}[x_1, \ldots, x_n]$ of degree d with $n \geq n(d)$ has a non-trivial p-adic zero.*

Theorem 3 (Birch, 1957 [2]). *For every odd degree d there is an integer $n(d)$ such every form $F(x_1, \ldots, x_n) \in \mathbb{Z}[x_1, \ldots, x_n]$ of degree d with $n \geq n(d)$ has a non-trivial zero in $\mathbb{Z}^n$.*

The proofs use multiply nested inductions, and produce values of $n(d)$ that grow in the same way as the Ackermann function.

One can also tackle the local-to-global problem via analytic methods, most notably the circle method. Under suitable circumstances this will show that

$$\#\{\mathbf{x} \in \mathbb{Z}^n : F(\mathbf{x}) = 0, \ \max_{i=1,\ldots,n} |x_i| \leq B\} = c_F B^{n-d} + o(B^{n-d}) \qquad (2)$$

as $B \to \infty$. Moreover the constant c_F is strictly positive providing that $F(x_1, \ldots, x_n) = 0$ has non-singular zeros in $\mathbb{R}^n$ and in each p-adic field. Under this condition we get an asymptotic formula for the number of solutions to $F(x_1, \ldots, x_n) = 0$ in a large cube. This provides a local-to-global result (with the proviso that we require our p-adic solutions to be non-singular).

A precise result of the above type is the following, which is contained in a theorem of Birch, from 1962 [3].

Theorem 4. *If the form F is non-singular, then* (2) *holds when* $n \geq 1 + (d-1)2^d$.

This gives a decent lower bound for n, but requires the form to be non-singular. Moreover it leaves open the p-adic question. For small d the lower bound on n can be improved. For $d = 2$ the theorem requires $n \geq 5$, but in fact (2) holds, in a suitably modified form, for $n = 3$ and 4 (Heath-Brown [9]). When $d = 3$ the original range $n \geq 17$ was improved to $n \geq 9$ by Hooley [11], in 1994. Finally, for $d = 4$, Browning and Heath-Brown, in work to appear [6], have shown that one can take $n \geq 41$, rather than $n \geq 49$.

Birch's theorem leads one to ask what one can expect for singular forms. When $d = 4$ one can consider

$$F(x_1, \ldots, x_n) = (x_1^2 - 2x_2^2)^2 + x_3^4 + \ldots + x_n^4.$$

It is easy to see that $F = 0$ has a real solution, and if $n \geq 18$ there is a p-adic solution for every p. However there is no non-trivial integer solution, no matter how large we take n. On the other hand, when $d = 3$ there is a non-trivial integral solution if n is large enough, as Birch's 1957 theorem shows. Davenport proved in 1963 [7] that $n \geq 16$ suffices, and Heath-Brown has shown in 2007 [10] that it is enough to take $n \geq 14$, for any cubic form over $\mathbb{Z}$.

We concentrate now on the p-adic problem. Artin asked if a form of degree d over $\mathbb{Z}$ has a non-trivial p-adic zero when $n \geq 1 + d^2$. This is implicit in Meyer's result for $d = 2$, and was proved by Lewis [13] (1952) for $d = 3$. However it is not known for any other d. An important approximation to Artin's Conjecture is given by the following result.

Theorem 5 (Ax and Kochen, 1965, [1]). *For every $d \in \mathbb{N}$ there is a $p_0(d)$ such that Artin's conjecture holds whenever $p \geq p_0(d)$.*

The proof uses Mathematical Logic, and is based on the fact that the analogue of Artin's conjecture is known for the fields $\mathbb{F}_p((t))$. A value for $p_0(d)$ was found by Brown [5] (1978):-

$$2^{2^{2^{2^{2^{2^{d^{11^{4d}}}}}}}} \ !$$

Here the "!" symbol is merely an exclamation mark, and not a factorial sign! Much smaller values have been found for certain small d. Leep and Yeomans [12] (1996) showed that one may take $p_0(5) = 47$, and indeed calculations by Heath-Brown (to appear) show that one can improve this to $p_0(5) = 17$. Wooley, in work also to appear, has shown that $p_0(7) = 883$ and $p_0(11) = 8053$ are admissible. Unfortunately the methods used here appear only to work for exponents d which cannot be written as a sum of composite numbers, that is to say, only for $d = 2, 3, 5, 7$ and 11.

Mathematical logicians have shown that the theory of p-adic fields is decidable. Thus for any give d and p there is, in principle, a decision procedure which will show whether or not Artin's Conjecture holds for forms of degree d over $\mathbb{Q}_p$. It follows that we can test the truth of Artin's Conjecture for any given degree d.

In spite of this evidence in favour of Artin's Conjecture it is now known to be false. Examples for $d = 4$ and $p = 2$ were given by Terjanian (1966 and 1980, [14, 15]), and we now have counter-examples for many values of d (though in every such example d is even). The simplest example is formed by setting

$$G(x_1, x_2, x_3) = x_1^4 + x_2^4 + x_3^4 - (x_1^2 x_2^2 + x_1^2 x_3^2 + x_2^2 x_3^2)\ x_1 x_2 x_3 (x_1 + x_2 + x_3)$$

and

$$
\begin{aligned}
F(x_1, \ldots, x_{18}) = &\ G(x_1, x_2, x_3) + G(x_4, x_5, x_6) + G(x_7, x_8, x_9) \\
&+ 4G(x_{10}, x_{11}, x_{12}) + 4G(x_{13}, x_{14}, x_{15}) \\
&+ 4G(x_{16}, x_{17}, x_{18}).
\end{aligned}
$$

One readily checks that $G(x_1, x_2, x_3) \equiv 1 \pmod{4}$ unless $2 | x_1, x_2, x_3$, and hence that $16 | F(x_1, \ldots, x_{18})$ implies $2 | x_1, \ldots, x_{18}$. This provides a counter-example with 18 variables, and Terjanian [15] produced a further version with 20 variables.

Given that Artin's conjecture is false, it is natural to ask just what one can say about the number $n(d)$, defined as the minimal n such that every form $F(x_1, \ldots, x_n) \in \mathbb{Z}[x_1, \ldots, x_n]$ of degree d has a non-trivial p-adic zero, for each p. Here we shall concentrate on quartics, this being the first open case. The theorem of Brauer [4] mentioned earlier shows that $n(d)$ exists, and Terjanian's counter-example shows that $n(4) \geq 21$. While Brauer's proof did not lead to feasible bounds for $n(d)$, more recent versions of the argument due to Schmidt, and particularly Wooley [16] (1998), are vastly more efficient. Wooley proves in general that

$$n(d) \leq d^{2^d},$$

and hence in particular that $n(4) \leq 4294967296$. Indeed, with only slightly more effort one can read off from Wooley's proof that in the particular case $d = 4$ one has $n(4) \leq 624293$. Such results are none the less very far from the value $n(4) = 17$ which Artin had predicted, or indeed from the best known lower bound $n(4) \geq 21$.

In the remainder of this lecture we describe how the upper bound for $n(4)$ may be improved, by a modification of the Brauer-Wooley method. Brauer's basic idea is that of "quasi-diagonalisation". It is not hard to show (Davenport and Lewis [8], 1963) that for every p and every d one can solve diagonal equations

$$c_1 x_1^d + \ldots + c_m x_m^d = 0$$

over $\mathbb{Z}_p$ as soon as $m > d^2$. If $F(x_1, \ldots, x_n)$ is a general integral form of degree d in n varaibles, we look for linearly independent vectors $\mathbf{e}_1, \ldots, \mathbf{e}_k \in \mathbb{Q}_p^n$ such that

$$F(\lambda_1 \mathbf{e}_1 + \ldots + \lambda_k \mathbf{e}_k) = c_1 \lambda_1^d + \ldots + c_k \lambda_k^d$$

for a suitable value of k. When $d = 2$ this is possible with $k = n$, that is to say we can diagonalise F, but in general we can only achieve this with some $k < n$. However, if we can find a suitable set of vectors $\mathbf{e}_1, \ldots, \mathbf{e}_k$ with $k > n^2$, then it is clear that an appropriate choice of the λ_i leads to a non-trivial p-adic zero of F.

The vectors $\mathbf{e}_i$ are found by induction on k, by solving simultaneous equations over $\mathbb{Q}_p$ involving forms of degree strictly less than d. Thus we also use induction on d, and information about the solvability of systems of forms. We illustrate the procedure for quartic forms. Suppose we have a value $N(r_1, r_2, r_3)$ such that one can solve a system of r_1 linear, r_2 quadratic, and r_3 cubic forms over any p-adic field, if we have $n > N(r_1, r_2, r_3)$ variables. Suppose also that we have suitable vectors $\mathbf{e}_1, \ldots, \mathbf{e}_k \in \mathbb{Q}_p^n$ and that we are looking for $\mathbf{e}_{k+1}$. We define bi-homogeneous polynomials $F_u(\mathbf{x}, \mathbf{y})$ by taking

$$F(t\mathbf{x} + \mathbf{y}) = \sum_{u=0}^{4} t^u F_u(\mathbf{x}, \mathbf{y}),$$

so that $F_u(\mathbf{x}, \mathbf{y})$ has degree u in $\mathbf{x}$ and degree $4 - u$ in $\mathbf{y}$. Let $T = \langle \mathbf{e}_1, \ldots, \mathbf{e}_k \rangle$, and let S be any direct complement of T in $\mathbb{Q}_p^n$. We then look for a solution vector $\mathbf{x}_0 \in S$ of the simultaneous equations

$$F_u(\mathbf{x}_0, \mathbf{y}) = 0 \quad \forall \mathbf{y} \in T, \quad \forall u \in \{1, 2, 3\}. \tag{3}$$

If we can find such an $\mathbf{x}_0$ then $\mathbf{e}_{k+1} = \mathbf{x}_0$ will have all the properties we require.

The system (3) contains infinitely many equations. We therefore write $\mathbf{y} \in T$ as

$$\mathbf{y} = \lambda_1 \mathbf{e}_1 + \ldots + \lambda_k \mathbf{e}_k,$$

and

$$F_u(\mathbf{x}, \mathbf{y}) = \sum_{\mathbf{d}} \lambda^{\mathbf{d}} F_{u,\mathbf{d}}(\mathbf{x}),$$

where

$$\lambda^{\mathbf{d}} = \lambda_1^{d_1} \ldots \lambda_k^{d_k}$$

runs over all monomials of degree $4 - u$, and $F_{u,\mathbf{d}}(\mathbf{x})$ is a form of degree u. Now, instead of the system (3), it suffices to have

$$F_{u,\mathbf{d}}(\mathbf{x}_0) = 0 \quad \forall \mathbf{d}, \quad \forall u \in \{1, 2, 3\}. \tag{4}$$

With $u = 1$ the number of monomials $\lambda^{\mathbf{d}}$ of degree 3 is $k(k+1)(k+2)/6$. Hence we have $k(k + 1)(k + 2)/6$ linear equations to solve. Similarly for $u = 2$ we find that the number of quadratics in the system (4) is $k(k + 1)/2$, while for $u = 3$ we get k cubic forms.

We therefore conclude that we may complete the induction step and find a suitable vector $\mathbf{e}_{k+1}$ providing that

$$\dim(S) = n - k > N\left(k, \frac{k(k+1)}{2}, \frac{k(k+1)(k+2)}{6}\right).$$

Since we needed to find $\mathbf{e}_1, \ldots, \mathbf{e}_{17}$ to estimate $n(4)$ we deduce that

$$n(4) \le 17 + N(16, 136, 816).$$

It is evident from the above argument that it would be good if we only had to use diagonal forms in 16 (say) variables, rather than 17. However, when $p = 5$ there are diagonal forms in 16 variables which have only the trivial 5-adic zero. An example is

$$F_0(x_1, \ldots, x_{16}) = x_1^4 + x_2^4 + x_3^4 + x_4^4 + 5(x_5^4 + x_6^4 + x_7^4 + x_8^4)$$

$$+ 25(x_9^4 + x_{10}^4 + x_{11}^4 + x_{12}^4)$$

$$+ 125(x_{13}^4 + x_{14}^4 + x_{15}^4 + x_{16}^4).$$

Thus it appears at first sight that one is forced to continue the induction as far as $\mathbf{e}_{17}$.

However there is a trick that can be used very effectively here. This is based on the fact that F_0 is (essentially) the only diagonal form which causes a problem. Suppose we have found $\mathbf{e}_1, \ldots, \mathbf{e}_{16}$ for which $F(\lambda_1\mathbf{e}_1 + \ldots + \lambda_{16}\mathbf{e}_{16})$ is diagonal. Then, unless the diagonal form is (essentially) the F_0 above, we will be able to produce a non-trivial p-adic zero of F by a suitable choice of $\lambda_1, \ldots, \lambda_{16}$. So suppose that

$$F(\lambda_1\mathbf{e}_1 + \ldots + \lambda_{16}\mathbf{e}_{16}) = F_0(\lambda_1, \ldots, \lambda_{16}).$$

We now define $\mathbf{e}'_{15} = \mathbf{e}_{15} + \mathbf{e}_{16}$ so that

$$
\begin{aligned}
F(\mu_1\mathbf{e}_1 + &\ldots + \mu_{14}\mathbf{e}_{14} + \mu_{15}\mathbf{e}'_{15}) \\
= \mu_1^4 &+ \mu_2^4 + \mu_3^4 + \mu_4^4 + 5(\mu_5^4 + \mu_6^4 + \mu_7^4 + \mu_8^4) \\
&+ 25(\mu_9^4 + \mu_{10}^4 + \mu_{11}^4 + \mu_{12}^4) + 125(\mu_{13}^4 + \mu_{14}^4 + 2\mu_{15}^4).
\end{aligned}
$$

If we now apply the induction process to find a further vector $\mathbf{e}'_{16}$ such that $F(\mu_1\mathbf{e}_1 + \ldots + \mu_{14}\mathbf{e}_{14} + \mu_{15}\mathbf{e}'_{15} + \mu_{16}\mathbf{e}'_{16})$ is diagonal, it must be of the shape

$$\mu_1^4 + \mu_2^4 + \mu_3^4 + \mu_4^4 + 5(\mu_5^4 + \mu_6^4 + \mu_7^4 + \mu_8^4) + 25(\mu_9^4 + \mu_{10}^4 + \mu_{11}^4 + \mu_{12}^4)$$

$$+ 125(\mu_{13}^4 + \mu_{14}^4 + 2\mu_{15}^4) + c\mu_{16}^4,$$

for some c. Thus it is not possible for this new diagonal form to be (essentially) equal to F_0, because of the coefficient 2 in front of μ_{15}^4. We have therefore produced a diagonal form which does have a non-trivial 5-adic zero, without needing to use the induction process to consider a 17-th basis vector.

It turns out that this process of contracting two basis vectors to produce a better diagonal form, and then adding a new vector, can be very useful in improving the number of variables required overall. However there are other ideas which we have not discussed here. This is currently the subject of ongoing research. However we state the following, which should be treated with caution. As the investigation proceeds the bounds may be improved; but equally it may happen that errors are detected and that the bounds get larger!

Provisional Theorem 1. A quartic form in n variables over $\mathbb{Q}_p$ has a non-trivial p-adic zero if

1. $n \geq 15387$, for $p = 2$;
2. $n \geq 609$, for $3 \leq p \leq 29$: and
3. $n \geq 131$, for $p \geq 31$.

We end with a few open questions.

Question 1. Does every cubic form over $\mathbb{Q}$, in $n \geq 10$ variables, have a non-trivial zero?

This very simple statement would be the natural analogue of Meyer's result for quadratic forms. It is already known for non-singular forms, and for singular forms when $n \geq 14$.

Question 2. Can one find any counter-examples to Artin's Conjecture with odd degree?

Perhaps a numerical search might find something.

Question 3. Are there any counter-examples to Artin's Conjecture for quartic forms with $p \neq 2$?

Again a numerical search might be useful.

Question 4. Does Artin's Conjecture hold for $d = 5$, for every prime?

This is certainly decidable in principle, but whether it is realistic to expect a computational proof with current technology is unclear.

References

[1] J. AX and S. KOCHEN, *Diophantine problems over local fields*, I, Amer. J. Math. **87** (1965), 605–630.

[2] B. J. BIRCH, *Homogeneous forms of odd degree in a large number of variables*, Mathematika **4** (1957), 102–105.

[3] B. J. BIRCH, *Forms in many variables*, Proc. Roy. Soc. Ser. A **265** (1961/1962), 245–263.

[4] R. BRAUER, *A note on systems of homogeneous algebraic equations*, Bull. Amer. Math. Soc. **51** (1945), 749–755.

[5] S. S. BROWN, *Bounds on transfer principles for algebraically closed and complete discretely valued fields*, Mem. Amer. Math. Soc. **15** (1978), no. 204, iv+92pp.

[6] T. D. BROWNING and D. R. HEATH-BROWN, *Rational points on quartic hypersurfaces*, to appear.

[7] H. DAVENPORT, *Cubic forms in sixteen variables*, Proc. Roy. Soc. Ser. A **272** (1963), 285–303.

[8] H. DAVENPORT and D. J. LEWIS, *Homogeneous additive equations*, Proc. Roy. Soc. Ser. A 274 (1963), 443–460.

[9] D. R. HEATH-BROWN, *A new form of the circle method, and its application to quadratic forms*, J. Reine Angew. Math. **481** (1996), 149–206.

[10] D. R. HEATH-BROWN, *Cubic forms in 14 variables*, Inventiones. Math. **170** (2007), 199–230.

[11] C. HOOLEY, *On nonary cubic forms*, J. Reine Angew. Math. **386** (1988), 32–98.

[12] D. B. LEEP and C. C. YEOMANS, *Quintic forms over p-adic fields*, J. Number Theory **57** (1996), 231–241.

[13] D. J. LEWIS, *Cubic homogeneous polynomials over p-adic number fields*, Ann. of Math. (2) **56** (1952), 473–478.

[14] G. TERJANIAN, *Un contre-exemple à une conjecture d'Artin*, C. R. Acad. Sci. Paris Sér. A-B **262** (1966), A612.

[15] G. TERJANIAN, *Formes p-adiques anisotropes*, J. Reine Angew. Math. **313** (1980), 217–220.

[16] T. D. WOOLEY, *On the local solubility of Diophantine systems*, Compositio Math. **111** (1998), 149–165.

PDE aspects of the Navier-Stokes equations

Vladimír Šverák

Abstract. In this lecture we will discuss mathematical aspects of the Navier-Stokes equations. We will recall some of the important open problems and mention a few recent results.

Consider a ball of radius R moving in an incompressible fluid of constant density ρ at constant velocity U. We expect that a force F is needed to keep the ball in motion (to overcome the "resistance of the medium"). The force is usually called the drag force. What is the formula for the drag force? This classical problem of Fluid Mechanics was considered already by Newton, who derived the formula

$$F = c\,\rho R^2 U^2 \,, \tag{1}$$

where c is a dimensionless constant. The formula (published in 1687) can be found in *Principia*, Corollary 1 of Theorem 30, Book II. From the modern point of view we can see that the formula is dictated by the dimensional analysis: the given expression for F is the only possible expression with the dimension of force which can be formed from the available data ρ, R, and U.

In 1752 d'Alembert published the well-known *Essai d'une nouvelle théorie de la résistance des fluides*, where he reached the surprising conclusion that in ideal fluids the drag force is

$$F = 0 \,. \tag{2}$$

This is known as d'Alembert's paradox.[1]

Supported in part by NSF Grant DMS-0457061

[1] The paradox has been well understood since the work of Prandtl in the early 1900s. The source of the paradox is the "ideal fluid" assumption, which is not satisfied for the usual fluids. Even very small internal friction in the fluid can have large effects when the fluid interacts with rigid boundaries. This is not captured by the ideal fluid model. See [23].

In 1851 Stokes brought viscosity into the considerations, and derived that for slowly moving objects one should have

$$F = 6\pi \nu \rho R U \,,\qquad (3)$$

where ν is the *kinematic viscosity* (which has dimension [Length]^2/[Time]) of the fluid.

The mathematical description of the fluid motion we use today is the same as the one used by Stokes, the Navier-Stokes equations:

$$\begin{aligned} u_t + u\nabla u + \tfrac{1}{\rho}\nabla p - \nu\Delta u &= 0 \\ \operatorname{div} u &= 0 \,, \end{aligned}\qquad (4)$$

where $u = (u_1(x,t), u_2(x,t), u_3(x,t))$ is the velocity of the fluid particle which is at position x at time t, and $p = p(x,t)$ is the pressure. For $\nu = 0$ the system (4) was derived in by Euler in 1757.

The natural boundary conditions for $\nu > 0$ is that $u = 0$ at the boundaries (in the local coordinates in which the boundary is at rest). For $\nu = 0$ the natural boundary condition is $u(x,t) \cdot n(x) = 0$, where $n(x)$ is the outward unit normal to the boundary.

The formulae of Stokes and d'Alembert are well understood in the context of PDE (4). For the Stokes formula one calculates (following Stokes) an explicit solution of the linearized problem. For d'Alembert's formula one assumes $\nu = 0$ and calculates that a steady-state solution of Euler's equations with the natural boundary conditions indeed leads to $F = 0$. (See, for example, [16].)

The formula of Newton is much more intriguing from the PDE point of view. Before we start its discussion, let us introduce an important dimensionless parameter of the above flow around a ball. We define the *Reynolds number* (introduced by Reynolds in 1880's) by

$$\mathrm{Re} = \frac{RU}{\nu} \,.\qquad (5)$$

Flows with the same Reynolds number are equivalent in the sense that the non-trivial scaling symmetries of the Navier-Stokes equations

$$u(x,t),\ p(x,t) \longrightarrow \lambda u(\lambda x, \lambda^2 t),\ \lambda^2 p(\lambda x, \lambda^2 t)\qquad (6)$$

can be used to map the situations with the same Reynolds number onto one another.[2]

---

[2] Here we of course have in mind flows defined in the exterior of balls.

From experiments we know that Newton's formula (1) is nearly correct once the Reynolds number is large (Re $\geq 10^6$ should be sufficient). From the point of view of PDEs this is remarkable, since ν plays a prominent role in the equation, and yet the force given by Newton's formula is independent of ν.

There is another remarkable classical experimental fact (discovered in the early 1900s by Prandtl and Eiffel): in our experiment described above, there is a certain Reynolds number Re_c, typically in the range $10^5 - 10^6$ (where the c in Newton's formula is not yet constant[3]), such that as we increase the velocity, the drag force will suddenly noticeably *decrease* as the Reynolds number crosses the critical value Re_c. This phenomenon is known as the *drag crisis* (see for example [16]), and (experimentally) it is related to a change of the geometry of the flow.

Are the above examples of fluid behavior described by the Navier-Stokes equations? The general belief is that this is indeed the case. However, strictly speaking, we do not really know, since the behavior is known only from experiments and not from computations or theoretical analyses of the equations. It is perhaps worth remarking that since one cannot really do experiments in two dimensions, we do not know if the two-dimensional Navier-Stokes exhibits similar behavior at large Reynolds numbers.

One should mention the following non-trivial issue which comes up in connection with the above problem of calculating the drag force. In our particular situation we want to solve the Navier-Stokes in the exterior domain $\Omega = \{x \in \mathbf{R}^3, \ |x| > R\}$ with the boundary conditions $u(x, t) = 0$ at the boundary $\partial\Omega = \{|x| = R\}$ and $u(x, t) \to U$ (where U is now considered as a vector) as $x \to \infty$. The Navier-Stokes equation (4) in Ω with these boundary conditions has various solutions. For example, one can consider axi-symmetric steady-state solutions. It can be proved that such solutions exist (this is essentially due to Leray). However, these solutions do not give the right drag force for the large Reynolds numbers. For example, it is likely (although it may not be known rigorously) that the drag force for these solutions approaches zero as the viscosity ν approaches zero, in contrast with what is observed in experiments. The reason is that the steady-state symmetric solutions are unstable, and the stable flows are neither symmetric nor time-independent. This has to be taken into account in numerical simulations. In reality the drag force F is a time average of the instantaneous force $F(t) =$

---

[3] The graphs of the dependence of c on Re obtained in experiments can be found in many Fluid Mechanics textbooks, see for example [16].

$\int_{\partial\Omega}[p(x,t)n(x)-2ve(x,t)n(x)]\,dx$, where $e(x,t)$ is the symmetric part of $\nabla u(x,t)$ and $n(x)$ is the unit normal.

Can the drag force be calculated numerically? With the best present-day computers, we cannot reliably solve the equations for Reynolds numbers exceeding 10^4. The reason is the appearance of various fine-scale structures, which make it difficult to resolve the details of the solutions. How much resolution is needed? There is a statistical-type theory of turbulence due to Kolmogorov and Onsager (see, for example, [16]) which is not based on the equations of motion, but rather on the assumption that certain statistics of the observed vector fields are invariant under natural scalings.[4] In the situation concerning the drag force, the assumed statistics implies that the drag force is independent of the viscosity (for large Reynolds numbers). The rough prediction from this theory is that the amount of computation needed to calculate the drag force at Reynolds number Re by "brute force" (*i.e.* by fully resolving the equations) is proportional to $\mathrm{Re}^{11/4}$. This rough prediction is probably too optimistic,[5] but it is still worth noticing that if we believe it and if we assume we could perhaps do Re $= 10^4$ today,[6] we need roughly 10^4 – fold increase in the computational power to calculate the air flow around tennis balls at speeds common in the game. To resolve the air flow around a car at realistic speeds (it can still be considered incompressible to a very good approximation), we would need roughly a 10^8 – fold increase in the computational power. Engineers of course need to calculate flows around cars (and airplanes) today, and to achieve this it is necessary to give up the hope of finding the full solution of the equations, and try instead to find some approximations. The art of finding such approximations is a large research area by itself, and there have been many partial successes. The main idea is that we do not need to resolve the solution fully, and that it should be possible to replace most of the unknown degrees of freedom by a suitable statistics. The famous open *problem of turbulence* can be thought of as the problem of finding a good algorithm which would do such a reduction to a relatively small number of variables in a reliable way in the general situation.

There is an additional problem one has to face. Namely, it is not known if (in dimension three) the Navier-Stokes equations (4) admit a smooth

---

[4] This is somewhat analogous to the situation in Statistical Mechanics where the laws are not really justified from the equations of motion, but are taken as new postulates.

[5] Among other things, it does not take into account the subtleties concerning the boundary layer.

[6] This would not be a calculation which could be done on a PC or a work station, we have in mind a really big computer.

solution which would describe the flows encountered in the drag force problem. This is known as the regularity problem. The simplest version of the problem is the following: does the initial-value problem for the system (4) in $\mathbf{R}^3 \times (0, \infty)$ with given initial data $u(x, 0) = u_0(x)$ have a smooth solution? The initial datum u_0 is assumed to be smooth and decay "sufficiently fast" to zero as $x \to \infty$.

This problem should be easier than the problem of turbulence, but it is still universally considered as a hard mathematical problem. In dimension two it has been solved a long time ago (by Leray in domains without boundaries and by Ladyzhenskaya in domains with boundaries), see for example [15]. By contrast, the problem of turbulence (as defined above) is open even in dimension two.

The reason why the regularity problem in dimension three is hard (at least for the present-day PDE techniques) can be understood as follows. There is only a limited number of "general" tools available for the analysis of PDEs and none of these techniques seem to be sufficient. No special mathematical structure has been discovered in the equations, and therefore the present-day theory has to treat them, to a large degree, as "the general case". At the same time, one does not expect regularity for all equations in this general class.

The available techniques include:

- Linear estimates;
- Perturbation Analysis, see for example [11, 13, 17];
- Energy methods, see, for example, [3, 17];
- "Scalar techniques", such as the maximum principle and other comparison principles, Harnack inequalities, De Giorgi, Nash-Moser, and Krylov-Safonov estimates, etc.;
- re-scaling and blow-up techniques, classification of entire solutions, etc., as pioneered by De Giorgi for minimal surfaces.

As in other areas of PDEs, ideas which can be traced back to the work of De Giorgi have played play an important role.

In the case of equation (4), we have a good understanding of the estimates for the linear part of the equation (which is called the Stokes system). We also have the energy identity

$$\int_{\mathbf{R}^3} \frac{1}{2} |u(x, t_2)|^2 \, dx + \int_{t_1}^{t_2} \int_{\mathbf{R}^3} \nu |\nabla u(x, t)|^2 \, dx dt = \int_{\mathbf{R}^3} \frac{1}{2} |u(x, t_1)|^2 \, dx \,,$$

$$(7)$$

which gives us the control of the quantity on the left-hand side ("parabolic energy"). We can now try to combine the energy estimate with the linear estimates and check if we can handle the non-linear term as a perturbation. This is possible in dimension two, but the energy estimate is too weak in dimension three. Two is the borderline dimension for this argument. (Sometimes the term "critical" is used in such borderline situations.) In dimension two we have in fact another quantity which plays to our advantage, the vorticity $\omega = \operatorname{curl} u$. This is a scalar (we are in dimension two!) which satisfies

$$\omega_t + u\nabla\omega = \nu\Delta\omega \ . \tag{8}$$

For this equation we have the maximum principle, and in the absence of boundaries it is easy to see that ω must be bounded. The boundedness of ω is more than enough for regularity. From this point of view the situation is "subcritical" – we have more than we need. However, the argument does not apply near the boundaries, where no "subcritical" argument is known, and one has to work with the energy estimate at the "critical" level.

In dimension three the situation is "super-critical" and all these arguments break down. This can also be understood heuristically in the following terms. The non-linear term in the equation can generate small length scales from large length scales (or, in Fourier terms, large frequencies from low frequencies). The linear part of the equation (which is dissipative) damps the small length scales. In the subcritical case the damping is stronger than the transfer from the longer length scales, and therefore the solution will stay in the realm of the finite length scale, which translates to regularity. In the supercritical case the damping may be insufficient, unless there is some extra mechanism which would slow down the transfer to the small scales. The critical case is the borderline.

It is perhaps interesting to note that when the large length scales (or low frequencies) become the focus, the situation is reversed. This happens for example when we study the behavior at infinity of the steady-state solutions of Navier-Stokes in exterior domains. The steady-state equations are subcritical in dimension two and three, and (local) regularity presents no problems. On the other hand, the study the solutions near ∞ is in some sense the study of the behavior at very large length scales (or low frequencies), and the subcritical case becomes more difficult. In fact, in both dimension two and three the exact behavior of the steady state solutions as $x \to \infty$ represents a difficult open problem,[7] see for

---

[7] The 2d problem is more difficult than the 3d problem in this case, as the 2d equation is "more subcritical".

example [1, 10, 14]. The issues become somewhat similar to the issues one has to face in the 3d time-dependent regularity problem, except that the problems are on the other end of the length scale spectrum, and are most likely easier.[8] There seems to be some kind of vague duality here.

In the above approach one uses only the energy estimate, estimates for the linear part, and some simple properties of the non-linear term, which are satisfied for many other equations. It is expected that the class of equations which share with Navier-Stokes the properties which have been used in the regularity theory so far contains some equations which allow the singularity formation in finite time from smooth data, see for example [21] and [22].

There has been some limited success with trying to find hidden scalar quantities in the equations, and to use the rescalings together with suitable blow-up procedures. Such techniques have been used for example to rule out self-similar singularities [20,25], certain type of axi-symmetric singularities [4,5,12], and to prove unexpected regularity results in the (super-critical) model case of the 5-dimensional steady-state Navier-Stokes [9, 28]. However, for the general time-dependent solutions is 3d no such quantities are known.

There has been a lot of research on conditions which are sufficient for regularity of solutions. After the well-known work of Leray, Prodi, Serrin and Ladyzhenskaya (see, for example, [15, 17, 24, 27]) this program has been further developed for example in [3, 6]. One interesting recent development in this direction is the result that the boundedness of the spatial L^3 - norm $||u(t)||_3 = \{\int_{\mathbf{R}^3} |u(x, t)|^3 \, dx\}^{1/3}$ of the solution (independently of t) is sufficient for regularity, see [8]. An interesting feature of the proof of the result is a somewhat unexpected connection to the control theory of parabolic equations.

The L^3 norm is special in that it is invariant under the scaling symmetry of the equation. It has the same dimension as the kinematic viscosity, and hence the quantity $||u(t)||_3/\nu$ is dimensionless. In some sense, this quantity can play the role of the Reynolds number in the absence natural length-scales.[9]

It should be mentioned that already in 1934 Leray proved that the 3d Navier-Stokes equations always admit global weak solutions. These are solutions which make sense even when singularities appear. It is known

---

[8] Nevertheless, still sufficiently hard to have remained open since the early papers of Leray on the subject in 1930s. See [10].

[9] The fluid occupying the whole space $\mathbf{R}^3$ is a good example.

that the set of possible singular points must be relatively small: its 1-dimensional parabolic Hausdorff measure has to be zero, see [3].

The main drawback of the theory of the weak solutions is that it is unknown whether they are unique.[10] This implies some difficulties for applications. For example, in the problem of the calculation of the drag force we mentioned in the beginning, it would not be trivial to define what the drag force is if the classes of solutions we deal with are not unique. Presumably one would have to find a suitable invariant measure on the set of all possible weak solutions and use some averaging process to define the drag force. It is not clear to me whether this has been done.

As interesting as the above mentioned results may be, they are inadequate for a real understanding the full 3d regularity problem. The key for the understanding of the problem might be in Euler's equation (the case $\nu = 0$). The regularity problem for Euler's equation is also open in dimension three (and the existence of global regular solutions is known in dimension two). One remarkable feature of Euler's equation is that the equation is completely canonical, there are no free parameters. In some sense it is really a geometric equation, and there is indeed a lot of beautiful geometry behind it, see for example [2, 7, 19]. So far there has not been much success in combining the PDE tools used for Navier-Stokes with the geometry which is behind Euler. For example, one of the main non-trivial mathematical facts about the solutions of Euler's equation, the Kelvin-Helmholtz law (see e. g. [26]), has not really found too much use in the regularity theory. It seems that Analysis and Geometry need to be brought together in some new way to make progress on these problems.

References

[1] C. J. AMICK, *On Leray's problem of steady Navier-Stokes flow past a body in the plane*, Acta Math. **161** (1988), 71–130.

[2] V. ARNOLD and B. KHESIN, "Topological Hydrodynamics", Springer, 2nd printing, 1999.

[3] L. CAFFARELLI, R.-V. KOHN and L. NIRENBERG, *Partial regularity of suitable weak solutions of the Navier-Stokes equations*, Comm. Pure Appl. Math. **35** (1982), 771–831.

[4] C.-C. CHEN, R. STRAIN, T.-P. TSAI and H.-T. YAU, *Lower bound on the blow-up rate of the axisymmetric Navier-Stokes equations*, Int. Math. Res. Not. IMRN 2008, no. 9.

---

[10] In fact, some mathematicians consider the problem of the uniqueness of the weak solutions to be more important than the regularity problem.

[5] C.-C. CHEN, R. STRAIN, T.-P. TSAI and H.-T. YAU, *Lower bound on the blow-up rate of the axisymmetric Navier-Stokes equations II*, preprint arXiv:0709.4230.

[6] P. CONSTANTIN and C. FEFFERMAN, *Direction of vorticity and the problem of global regularity for the Navier-Stokes equations*, Indiana Univ. Math. J. **42** (1993), 775–789.

[7] D. EBIN and J. MARSDEN, *Groups of diffeomorphisms and the notion of an incompressible fluid*, Ann. of Math. (2) **92** (1970), 102–163.

[8] L. ESCAURIAZA, G. SEREGIN and V. ŠVERÁK, $L_{3,\infty}$-*Solutions to the Navier-Stokes equations and backward uniqueness*, Uspekhi Matematicheskih Nauk **58** 2 (350), 3–44. English translation in Russian Mathematical Surveys, **58** (2003), 211–250.

[9] J. FREHSE and M. RŮŽIČKA, *Regularity for the stationary Navier-Stokes equations in bounded domains*, Arch. Rational Mech. Anal. **128** (1994), 361–380.

[10] G. P. GALDI, "An Introduction to the Mathematical Theory of the Navier-Stokes Equations", Volumes I and II, Springer, 1994.

[11] T. KATO, *Strong L^p-solutions of the Navier-Stokes equation in $\mathbb{R}^m$*, *with applications to weak solutions*, Math. Z. **187** (1984), 471–480.

[12] G. KOCH, N. NADIRASHVILI, G. SEREGIN and V. SVERAK, *Liouville theorems for the Navier-Stokes equations and applications*, to appear in Acta math., see also arXiv:0709.3599.

[13] H. KOCH and D. TATARU, *Well-posedness for the Navier-Stokes equations*, Adv. Math. **157** (2001), 22–35.

[14] *On the large-distance asymptotics of steady state solutions of the Navier-Stokes equations in 3D exterior domains*, arXiv:0711.0560.

[15] O. A. LADYZHENSKAYA, "The mathematical theory of viscous incompressible flow", Revised English edition, Gordon and Breach Science Publishers, New York-London 1963.

[16] L. D. LANDAU and E. M. LIFSCHITZ, 'Fluid Mechanics", second edition, Butterworth-Heinemann, 2000 paperback reprinting.

[17] J. LERAY, *Sur le mouvement d'un liquide visqueux emplissant l'espace*, Acta Math. **63** (1934), 193–248.

[18] F. H. LIN, *A new proof of the Caffarelli-Kohn-Nirenberg theorem*, Comm. Pure Appl. Math., **51** (1998), 241–257.

[19] A. MAJDA and A. BERTOZZI,"Vorticity and Incompressible Flow", Cambridge Texts in Applied Mathematics, Vol. 27, Cambridge University Press, Cambridge, 2002.

[20] J. NEČAS, M. RŮŽIČKA and V. ŠVERÁK, *On Leray's self-similar solutions of the Navier-Stokes equations*, Acta Math. **176** (1996), 283–294.

[21] P. PLECHÁČ and V. ŠVERÁK, *On self-similar singular solutions of the complex Ginzburg-Landau equation*, Comm. Pure Appl. Math. **54** (2001), 1215–1242.

[22] P. PLECHÁČ and V. ŠVERÁK, *Singular and regular solutions of a nonlinear parabolic system*, Nonlinearity **16** (2003), 2083–2097.

[23] L. PRANDTL, *Über Flüssigkeitsbewegung bei sehr kleiner Reibung*, Proceedings of the Third International Congress of Mathematicians, Heidelberg 1904. Teubner, Leipzig, 1905.

[24] G. PRODI, *Un teorema di unicità per le equazioni di Navier-Stokes*, Ann. Mat. Pura Appl. (4) **48** (1959), 173–182.

[25] T.-P. TSAI, *On Leray's Self-Similar Solutions of the Navier-Stokes Equations Satisfying Local Energy Estimates*, Arch. Rational Mech. Anal. **143** (1998), 29–51.

[26] P. G. SAFFMAN, "Vortex Dynamics", Cambridge University Press, 1992.

[27] J. SERRIN, *On the interior regularity of weak solutions of the Navier-Stokes equations*, Arch. Rational Mech. Anal. **9** (1962), 187–195.

[28] M. STRUWE, *Regular solutions of the stationary Navier-Stokes equations on* $\mathbf{R}^5$, Math. Ann. **302** (1995), 719–741.

The canonical ring is finitely generated

Christopher D. Hacon

1. Introduction

Let X be a smooth complex projective variety so that X is a subset of $\mathbb{P}^N$ cut out by finitely many homogeneous polynomials $P_i \in \mathbb{C}[z_0, \ldots, z_N]$. The canonical bundle of X is denoted by ω_X so that for all $m \geq 0$ sections $s \in H^0(X, \omega_X^{\otimes m})$ may be locally written as $f \cdot (dx_1 \wedge \ldots \wedge dx_n)^{\otimes m}$ where f is a holomorphic function and $x_1, \ldots, x_n$ are local parameters on X. The vector spaces $H^0(X, \omega_X^{\otimes m})$ give rise to the canonical ring

$$R(\omega_X) := \bigoplus_{m \geq 0} H^0(X, \omega_X^{\otimes m}).$$

This ring is of fundamental importance in the study of the birational geometry of higher dimensional varieties. Recall that if X and X' are birational (*i.e.* they have isomorphic open subsets) then $H^0(X, \omega_X^{\otimes m}) \cong H^0(X', \omega_{X'}^{\otimes m})$. In particular $R(\omega_X)$ is a birational invariant of X. The purpose of this note is to give an overview of recent results in higher dimensional birational algebraic geometry that lead to the proof of the following:

Theorem 1.1. *Let X be a smooth complex projective variety. Then the canonical ring $R(\omega_X)$ is finitely generated.*

It should be noted that there are two announced proofs of this result. We will illustrate the approach of [2] and [1] which is based on the ideas of the minimal model program and in particular on ideas of V. Shokurov [6]. This approach uses the methods of higher dimensional birational geometry and has the pleasant feature that it also allows us to prove many important results on the birational geometry of higher dimensional complex projective varieties such as the existence of flips and (under favorable, but not too restrictive conditions) the termination of certain sequences of flips and hence the existence of minimal models.

As mentioned above, there is another proof due to Y.-T. Siu [7]. This is completely independent and is based on analytic methods. In this paper we will not discuss any of the details of the analytic approach.

We now recall some notation. $\mathbb{P}^n$ will denote n-dimensional complex projective space, *i.e.* a compactification of $\mathbb{C}^n$ obtained by adding a hyperplane (a copy of $\mathbb{P}^{n-1}$). A complex projective variety in $\mathbb{P}^n$ is given by the common zeroes of a finite set of homogeneous polynomials $P_1, \ldots, P_r \in \mathbb{C}[z_0, \ldots, z_n]$. Given a line bundle L on a variety X, $H^0(X, L)$ denotes the complex vector space of global sections of L. In particular $\mathcal{O}_{\mathbb{P}^n}(1)$ denotes the hyperplane line bundle so that $H^0(\mathbb{P}^n, \mathcal{O}_{\mathbb{P}^n}(r))$ may be identified with the homogeneous polynomials in $\mathbb{C}[z_0, \ldots, z_n]$ of degree r. There is a rational map $\phi_L : X \dashrightarrow \mathbb{P}^N = \mathbb{P}H^0(X, L)$ defined as follows: let $s_0, \ldots, s_N$ be a basis of $H^0(X, L)$ and $x \in X$, then $\phi_L(x) = [s_0(x), \ldots, s_N(x)]$. Note that ϕ_L is undefined at any point in the base locus of L.

For any line bundle L we let $R(L)$ be the graded ring given by $\bigoplus_{m \geq 0} H^0(X, L^{\otimes m})$. We remark that if k is a positive integer, then the ring $R(L)$ is finitely generated if and only if so is $R(L^{\otimes k})$ The number $\kappa(L) := \text{tr. deg. } R(L) - 1$ is the *Kodaira dimension* of X.

For any line bundle L and any curve $C \subset X$, we may define $L \cdot C = \deg(L|_C)$, and for any morphism $f : Y \to X$ we have a line bundle on Y given by f^*L. Similarly definitions also hold for any formal linear combination of line bundles $L = \sum r_i L_i$ where $r_i \in \mathbb{R}$.

For any birational morphism $f : Y \to X$, $\text{Exc}(f)$ denotes the exceptional set, so that $Y - \text{Exc}(f)$ is the biggest open subset of Y where f restricts to an isomorphism.

ACKNOWLEDGEMENTS. The author was partially supported by NSF grant 0456363 and by an AMS centennial scholarship. He would also like to thank the Scuola Normale Superiore di Pisa for its hospitality.

2. Geometry of curves

Curves are smooth complex projective varieties of dimension 1 also known as Riemann surfaces. These are topologically classified by their genus $g := \dim H^0(X, \omega_X)$ (recall that in this case $\omega_X = \Omega_X^1 = T_X^\vee$ is the cotangent bundle and so g is just the number of linearly independent global holomorphic 1-forms). We can divide surfaces in to three rough classes:

(1) $g = 0$. In this case there is only one possibility: $\mathbb{P}^1$. We have that $\omega_X = \mathcal{O}_{\mathbb{P}^1}(-2)$ so that the elements of $H^0(\mathbb{P}^1, \omega_X^{\otimes m}) \cong H^0(\mathbb{P}^1, \mathcal{O}_{\mathbb{P}^1}(-2m))$

correspond to homogeneous polynomials of degree $-2m$ and hence they are all 0. In particular $R(X) \cong \mathbb{C}$.

(2) $g = 1$. In this case there is a 1-parameter family of elliptic curves. We have that $\omega_X = \mathcal{O}_X$ and so $R(X) \cong \mathbb{C}[t]$.

(3) $g \geq 2$. In this case there is a $(3g-3)$-parameter family of such curves. One can consider sections of the line bundle $L := \omega_X^{\otimes 3}$. It is known that the sections of L define an embedding $\phi_L : X \to \mathbb{P}^{5g-6} = \mathbb{P}H^0(L)$. One has that $H^0(X, \omega_X^{\otimes 3m}) \cong H^0(X, \mathcal{O}_{\mathbb{P}^{5g-6}}(m)|_X)$. For any $m \gg 0$ the homomorphisms $H^0(\mathbb{P}^{5g-6}, \mathcal{O}_{\mathbb{P}^{5g-6}}(m)) \to H^0(X, \mathcal{O}_{\mathbb{P}^{5g-6}}(m)|_X)$ are surjective. Since $R(\mathcal{O}_{\mathbb{P}^{5g-6}}(1))$ is finitely generated, then so is $R(\omega_X)$.

The natural question is if one can then generalize these results to higher dimensions.

3. Geometry of Surfaces

The first problem that one encounters in the classification of surfaces (*i.e.* complex projective varieties of dimension 2) is that given any surface X, one can produce a new surface by blowing up a point $x \in X$. This produces a morphism $f : X' = \mathrm{bl}_x X \to X$ which is an isomorphism over $X - x$ and replaces the point x by a -1 curve *i.e.* a rational curve $E \cong \mathbb{P}^1$ such that $E^2 = -1$. The points of E correspond to the tangent directions at x. It is then reasonable to attempt to classify surfaces modulo birational isomorphism, so that two surfaces (or higher dimensional varieties) are equivalent if they have isomorphic open subsets. It is known that two surfaces are birational if they are isomorphic after a finite sequence of blow ups. By Castelnuovo's criterion one may blow down any -1 curve, so that one sees that any surface is birational to a minimal surface *i.e.* a surface that contains no -1 curves.

The second problem is what invariant should replace the genus in higher dimension. One option would be to take the topological type of X, but this notion is too rigid and in particular it is not invariant under birational maps. It turns out that a better choice is given by the vector spaces $H^0(X, \omega_X^{\otimes m})$ for $m > 0$. An element of $H^0(X, \omega_X^{\otimes m})$ is called a global m-th pluricanonical form. It can be locally written as $f \cdot (dz_1 \wedge dz_2)^{\otimes m}$ where f is a holomorphic function and z_1 and z_2 are local parameters. As mentioned above, we have a graded ring known as the canonical ring $R(\omega_X) = \bigoplus_{m>0} H^0(X, \omega_X^{\otimes m})$ which is a birational invariant of X. One may also define a coarser invariant, known as the Kodaira dimension of X given by

$$\kappa(X) = \mathrm{tr.\,deg.} R(\omega_X) - 1 \in \{-1, 0, 1, \ldots, \dim X\}.$$

We say that X is of general type if $\kappa(X) = \dim(X)$. For surfaces, we then have $\kappa(X) \in \{-1, 0, 1, 2\}$.

The Enriques-Iitaka classification of surfaces may be described as follows.

(1) $\kappa(X) = -1$: X is covered by rational curves (in fact X is birational to $C \times \mathbb{P}^1$ for some curve C of genus $g(C) = \dim H^0(\Omega_X^1)$). Therefore $H^0(X, \omega_X^{\otimes m}) \cong 0$ for all $m > 0$ and so the canonical ring $R(\omega_X)$ is isomorphic to $\mathbb{C}$.

Note that if $\kappa(X) = -1$, then X has many different minimal surfaces (but their relationships are well understood). On the other hand, if $\kappa(X) \geq 0$, it is known that X has a unique minimal surface say X'.

(2) $\kappa(X) = 0$: X is birational to a unique minimal surface X' which is in one of four well understood classes of surfaces (abelian, K3, Enriques and bielliptic surfaces). One has that $\omega_{X'}^{\otimes 12} \cong \mathcal{O}_{X'}$. The canonical ring $R(\omega_X)$ is isomorphic to $\mathbb{C}[t]$.

(3) $\kappa(X) = 1$: X is covered by elliptic curves. In fact X is birational to a unique minimal surface X' which admits a morphism $f : X' \to C$ such that $\omega_{X'}^{\otimes 12} = f^*L$ for some line bundle L of positive degree on C. It then follows that the canonical ring $R(\omega_X)$ is finitely generated since $R(\omega_X^{\otimes 12}) \cong R(L)$ is finitely generated.

(4) $\kappa(X) = 2$: X is birational to a unique minimal surface X' such that $\omega_{X'}$ is nef (i.e. $\deg(\omega_{X'}|_C) \geq 0$ for any curve $C \subset X'$). $H^0(\omega_{X'}^{\otimes 5})$ defines a birational morphism $f : X' \to \mathbb{P}^N = \mathbb{P}H^0(\omega_{X'}^{\otimes 5})$ which contracts all rational curves $E \cong \mathbb{P}^1$ with $E^2 = -2$ to a DuVal singularity. We therefore have that $\omega_{X'}^{\otimes 5} \cong f^*\mathcal{O}_{\mathbb{P}^N}(1)$ so that $R(\omega_X)$ is finitely generated and $f(X') \cong \operatorname{Proj} R(\omega_X)$.

4. Geometry of Threefolds

One would like to generalize the above classification to the case of threefolds. This is possible but there are several new features which made the problem extremely difficult. The classification was achieved by work of Kawamata, Kollár, Mori, Reid, Shokurov and others which culminated in Mori's construction of flips [5]. The upshot is the following.

Theorem 4.1. *Let X be a smooth complex projective 3-fold.*

(1) *If $\kappa(X) = -1$, then X is covered by rational curves.*
(2) *If $\kappa(X) \geq 0$, then X has a minimal model.*

In all cases the canonical ring $R(\omega_X)$ is finitely generated.

It is important to notice that the minimal model is not unique and may have mild singularities. We must in fact allow terminal singularities.

These are mild singularities in particular they are rational singularities that occur in codimension ≥ 3 and by definition $\omega_X^{\otimes m}$ is a line bundle for some $m > 0$ so that we may still define $\omega_X \cdot C = \frac{1}{m} \deg(\omega_X^{\otimes m}|_C)$. In dimension 3 these singularities are classified, but in higher dimension they are somewhat more mysterious (but still well behaved).

There is also an explicit procedure for constructing a minimal model known as the minimal model program. To run a minimal model program, one starts with a terminal complex projective variety X. If ω_X is nef, then we are done. Otherwise, let

$$N_1(X) = \left\{ \sum c_i C_i | c_i \in \mathbb{R}, \ C_i \text{ is a curve in } X \right\} / \equiv$$

where $C \equiv D$ (that is the curves C and D are numerically equivalent) if $(C-D) \cdot L = 0$ for any line bundle L on X. We let $\rho(X) = \dim_{\mathbb{R}}(N_1(X))$. Let $\overline{NE}(X)$ be the closure of the quotient of the cone of effective cones on X. If ω_X is not nef, then by the Cone Theorem there is an ample line bundle A and a rational number $a > 0$ such that $\omega_X \otimes A^{\otimes a}$ is nef and a unique *negative extremal ray* $R = \mathbb{R}^+[C]$ where C is curve in X and $(\omega_X \otimes A^{\otimes a}) \cdot C' = 0$ if and only if $[C'] \in R$. By the Base Point Free Theorem, there is then a morphism $f : X \to Z$ (surjective with connected fibers) such that for any curve $D \subset X$, $f_* D = 0$ if and only if $[D] \in R$. There are several cases to consider.

(1) If $\dim(Z) < \dim(X)$, then $f : X \to Z$ is called a *Mori fiber space*. It has the following properties: $\rho(X) - \rho(Z) = 1$; $\omega_X \cdot C < 0$ for any curve contracted by f; the fibers of f are covered by rational curves (in fact rationally connected). This gives a clear geometric reason why in this case $\kappa(X) = -1$ as ω_X has negative degree on this covering family of rational curves and so any element of $H^0(X, \omega_X^{\otimes m})$ must vanish on this covering family of rational curves and hence must be 0.

(2) If $\dim(Z) = \dim(X)$ and $\dim(\text{Exc}(f)) = \dim(Z) - 1$, then f is a *divisorial contraction*. In this case Z also has terminal singularities and we may simply replace X by Z. This is the analog of the contraction of a -1 curve in the surface case. We have $\rho(Z) = \rho(X) - 1$ and hence this process can be repeated only finitely many times.

(3) If $\dim(Z) = \dim(X)$ and $\dim(\text{Exc}(f)) < \dim(Z) - 1$, then f is a *small contraction*. In this case Z does not have terminal singularities. In fact $\omega_X^{\otimes m}$ is not a line bundle for any $m > 0$ and so one cannot make sense of the intersection number $\omega_X \cdot C$ of some curves $C \subset X$. The (very bold!) solution is then to replace X by its flip. The flip of $f : X \to Z$ is a morphism $f^+ : X^+ \to Z$ such that X is isomorphic to X^+ outside a codimension ≥ 2 subset, $\rho(X) = \rho(X^+) = \rho(Z) + 1$

and $\omega_{X^+} \cdot C > 0$ for any curve $C \subset X^+$ contracted by f^+. Therefore a flip can be thought of a codimension 2 surgery which replaces some ω_X negative curves by ω_{X^+} positive curves. Since our goal is to arrive to a minimal model (or to a Mori fiber space), this would seem to be a step in the right direction. The good news is that if the flip exists, then it is uniquely defined by the formula

$$X^+ = \text{Proj}_Z \bigoplus_{m \geq 0} f_*(\omega_X^{\otimes m})$$

and it has mild singularities. There are two items of bad news:

First of all it is very hard to prove the existence of X^+. In fact (assuming that Z is affine) this is equivalent to showing that $R(\omega_X)$ is finitely generated as an $\mathcal{O}_Z$ module. At first glance this would seem hopeless as it is equivalent to one of the original motivating problems: to show that $R(\omega_X)$ is finitely generated. Upon further reflection, one notices that as we want to only show that $R(\omega_X)$ is finitely generated over Z, the problem might be more accessible, especially in view of the fact that $\dim(X) = \dim(Z)$. At any rate, Mori solved the problem in the most geometric (but hardest) possible way: he classified all possible flipping contractions $f : X \to Z$ of this type and then constructed the corresponding flip X^+.

Secondly, one must show that any sequence of flips terminates. Luckily in dimension 3 this is not too difficult.

5. Minimal Model Program for log pairs

The Minimal Model Program is expected to work in the more general setting of log pairs. We let K_X denote a canonical divisor *i.e.* a divisor corresponding to ω_X. A *log pair* (X, B) consists of a normal variety X and a $\mathbb{R}$-divisor $B = \sum b_i B_i$ (*i.e.* $b_i \in \mathbb{R}$ and B_i are irreducible codimension 1 subvarieties) and $K_X + B$ is $\mathbb{R}$-Cartier (so that you may think of it as a formal linear combination of line bundles with real coefficients). Therefore, it still makes sense to consider pull-backs $f^*(K_X + B)$ and to intersect $K_X + B$ with curves $C \subset X$. In particular we can ask the question *is (X, B) a minimal model?* *i.e.* is $(K_X + B) \cdot C \geq 0$ for any curve $C \subset X$.

Just as in the case with no boundary (*i.e.* $B = 0$), we hope to find a birational map $\phi : X \dashrightarrow Y$ consisting of flips and divisorial contractions such that either

(1) $(Y, \phi_* B)$ is a minimal model (*i.e.* $K_Y + \phi_* B$ is nef), or

(2) $(Y, \phi_* B)$ is a Mori fiber space (*i.e.* there is a surjective morphism with connected fibers $f : Y \to S$ such that $\rho(Y) - \rho(S) = 1$ and $-(K_Y + \phi_* B) \cdot C < 0$ for any curve C contracted by f).

To achieve this we must require that the pair (X, B) have mild singularities. It turns out that one should require that (X, B) have kawamata log terminal singularities, so that if we have $f : X' \to X$ a morphism from a smooth variety X' such that the components of the transform of B and of the exceptional divisor are smooth and transverse, then we may write $f^*(K_X + B) = K_{X'} + B'$ where $B' = \sum b'_i B'_i$, $f_* B' = B$ and $b'_i < 1$ for all i.

The Cone Theorem is known to hold for kawamata log terminal pairs in any dimension. Therefore the main questions to answer are:

Question 5.1. Let (X, B) be a kawamata log terminal pair.

(1) Do flips exist?
(2) Is any given sequence of flips finite?

A positive answer to these questions would then allow us to construct minimal models in all dimension and hence show that pluricanonical rings are finitely generated.

6. Higher dimensional varieties

Using ideas of Shokurov, in [1] it is shown that:

Theorem 6.1. *Let (X, Δ) be a kawamata log terminal pair such that one of the following holds:*

(1) $\kappa(K_X + B) = \dim X$, *or*
(2) $\kappa(B) = \dim X$, *or*
(3) (X, B) *is not pseudo-effective, (i.e. for some ample divisor A and some $0 < \epsilon \ll 1$, we have $\kappa(K_X + B + \epsilon A) = -1$).*

Then there exists a finite sequence of flips and divisorial contractions $\phi : X \dashrightarrow Y$ such that $(Y, \phi_* B)$ is either a minimal model or is a Mori fiber space.

Corollary 6.2. *If X is a smooth complex projective variety, then its canonical ring $R(K_X)$ is finitely generated.*

Proof. This follows from a result of Mori and Fujino [4] according to which it suffices to prove finite generation for any kawamata log terminal pair (Y, B) with $\kappa(K_Y + B) = \dim Y$.   □

We remark that we prove that flips of kawamata log terminal pairs exist in all dimensions but we do not show that sequences of flips terminate. What we show is that under the right hypothesis, there exists a carefully chosen sequence of flips that terminates. These sequences of flips are know as flips with scaling. To explain this, suppose that we have a kawamata log terminal pair (X, B), then we may choose an ample divisor A and a number $0 < \tau_0 \le 1$ such that $K_X + B + \tau_0 A$ is ample (and hence nef). Then we let

$$\tau_1 := \inf\{t > 0 | K_X + B + tA \text{ is nef}\}.$$

If $\tau_1 = 0$, we are done as $K_X + B$ is nef. Otherwise there is a $K_X + B$ negative extremal ray $R = \mathbb{R}^+[C]$ such that $(K_X + B + \tau_1 A) \cdot C = 0$. If R induces a Mori fiber space, we are also done. Otherwise, we perform the corresponding flip or divisorial contraction say $\phi_1 : X \dashrightarrow X_1$. Since $(K_X + B + \tau_1 A) \cdot C = 0$ it follows that also $K_{X_1} + (\phi_1)_* B + \tau_1 (\phi_1)_* A$ is nef. We may therefore repeat this procedure. Proceeding in this way, we obtain a sequence of flips and divisorial contractions $\phi_i : X_{i-1} \dashrightarrow X_i$ and numbers $0 \le \tau_{i-1} \le \tau_i \le 1$ such that $K_{X_i} + B_i + \tau_i A_i$ is nef. This sequence ends if $\tau_n = 0$ or if we obtain a Mori fiber space. Otherwise we have an infinite sequence of minimal models $(X_i, B_i + \tau_i A_i)$.

The key idea is that if $\kappa(B) = \dim X$, then by using a compactness argument, we can show that there are only finitely many distinct minimal models for $(X, B + tA)$ where $0 \le t \le 1$.

In order to show that flips exist, we use Shokurov's so called reduction to PL-flips. The PL here stands for pre-limiting. Recall that given a flipping contraction $f : X \to Z$, to construct the flip of X, it suffices to show that

$$R(K_X + B) = \bigoplus_{m \ge 0} H^0(X, \mathcal{O}_X(m(K_X + B)))$$

is finitely generated as an $\mathcal{O}_Z$ algebra (here we are assuming for simplicity that Z is affine). The main idea is that after the reduction to PL-flips, we may assume that $B = B_0 + \sum b_i B_i$ where $b_i \in \mathbb{Q}$ and $0 \le b_i < 1$ (more precisely the pair (X, B) is purely log terminal so that if $v : X' \to X$ is a birational map and $K_{X'} + B' = v^*(K_X + B)$, then all coefficients of B' are < 1 except for the coefficient of the transform of B_0 which equals 1). We may also assume that a further technical condition holds, namely that for some positive number q, we have $(K_X + B - q B_0) \cdot C = 0$ for all curves $C \subset X$ contracted by $f : X \to Z$.

We then let $S := B_0$ and we look at the restriction map

$$\Psi : \bigoplus_{m \ge 0} H^0(X, \mathcal{O}_X(m(K_X + B))) \to \bigoplus_{m \ge 0} H^0(S, \mathcal{O}_S(m(K_S + B_S))).$$

From the definition of purely log terminal singularity, one can see that the pair (S, B_S) is kawamata log terminal. The idea is that the kernel of this map of graded rings is essentially a principal ideal so that in order to show that $R(K_X + B)$ is finitely generated, it suffices to show that $\text{Im}(\Psi)$ is finitely generated. Therefore if Ψ is surjective, we can then conclude by induction on the dimension. Unluckily this is too much to expect. However, in [2], we show that after replacing X and S by suitable birational models, there is a divisor $0 \leq \Theta \leq B_S$ such that

$$\text{Im}(\Psi) = \bigoplus_{m>0} H^0(S, \mathcal{O}_S(m(K_S + \Theta))).$$

Note that (S, Θ) is also a kawamata log terminal pair and so if the coefficients of Θ are rational we are once again done by induction on the dimension. It should be remarked however that Θ is obtained by a limiting procedure and hence is a priori only an $\mathbb{R}$-divisor. This problem can be addressed (as already observed by Shokurov) by using techniques of diophantine approximation.

References

[1] C. BIRKAR, P. CASCINI, C. D. HACON and J. M^{c}KERNAN, *Existence of minimal models for varieties of log general type*, ArXiv:math.AG/0610203.

[2] C. D. HACON and J. M^{c}KERNAN, *Extension Theorems and the existence of Flips*, In: "Flips for 3-folds and 4-folds", A. Corti (ed.), Oxford Lecture Series in Mathematics and its Applications **35**.

[3] J. KOLLÁR and S. MORI, "Birational Geometry of Algebraic Varieties", With the collaboration of C. H. Clemens and A. Corti, Translated from the 1998 Japanese original, Cambridge Tracts in Mathematics, Vol. 134, Cambridge University Press, Cambridge, 1998.

[4] O. FUJINO and S. MORI, *A canonical bundle formula*, J. Differential Geom. **56** (2000), 167–188.

[5] S. MORI, *Flip theorem and the existence of minimal models for 3-folds*, J. Amer. Math. Soc. **1** (1988), 117–253.

[6] V. V. SHOKUROV, *Prelimiting flips*, Tr. Mat. Inst. Steklova **240** (2003), Biratsion. Geom. Linein. Sist. Konechno Porozhdennye Algebry, 82–219; translation in Proc. Steklov Inst. Math. 2003, no. 1 (240), 75–213.

[7] Y.-T. SIU, "A General Non-Vanishing Theorem and an Analytic Proof of the Finite Generation of the Canonical Ring", math. AG/0610740.

Elliptic curves and Iwasawa theory

John Coates

Introduction

The earliest recorded European work on the arithmetic of elliptic curves occurs in Leonardo Pisano's book "Liber Quadratorum", published in the 13th century. Thus the oldest elliptic curve known to mankind in Europe

$$y^2 = x^3 - x$$

has a close association with Pisa, and it is a pleasure to give this lecture in Pisa on some of the subsequent developments in the subject.

1. Classical descent theory

An elliptic curve E over $\mathbb{Q}$ is a curve of genus 1 defined over $\mathbb{Q}$, endowed with a $\mathbb{Q}$-rational point σ, which is the origin of the group law. Let $E(\mathbb{Q})$ be the group of $\mathbb{Q}$-rational points on E. By brilliantly generalizing Fermat's notion of *infinite descent*, Mordell proved the following theorem in 1923:

Theorem 1.1 (Mordell). $E(\mathbb{Q})$ *is a finitely generated abelian group.*

We define g_E to be the rank of $E(\mathbb{Q})$. In *numerical* examples, we can today usually determine g_E.

Example 1.2. $E : y^2 = x^3 + 14x.$

Then $g_E = 2$ and $E(\mathbb{Q})$ is generated by $(2, 6)$, $(\frac{1}{4}, \frac{15}{8})$, and the point $(0, 0)$ of order 2.

The reason why the subject remains deeply interesting today is that no *theoretical* algorithm has ever been *proven* for determining the invariant g_E. Here is a typical example of an unsolved problem which can be traced back to Arab manuscripts at least a thousand years ago.

Conjecture 1.3. *Let N be a square free positive integer with $N \equiv 5, 6, 7$ mod 8. Then we always have $g_E \geq 1$ for the curve*

$$E : y^2 = x^3 - N^2 x \,. \tag{1.1}$$

Of course, this conjecture is known to be true for many specific values of N, and the cases $N = 5, 6, 7$ go back deep into antiquity in Asia.

Why are these problems so difficult theoretically? The essential reason is that the Mordell-Fermat argument of infinite descent is really a *cohomological argument* about *phantom points* on E. We sketch it. We pick a prime number p, and let $\mathbb{Q}_p$, $\mathbb{Z}_p$ denote the field of p-adic numbers and the ring of p-adic integers.

Definition 1.4. E_{p^∞} = Galois module of all p-power division points on E.

The proof proceeds by constructing a canonical injection

$$i_{E,p} : E(\mathbb{Q}) \otimes \mathbb{Q}_p/\mathbb{Z}_p \to \mathcal{S}_{E,p}$$

where $\mathcal{S}_{E,p}$ (= Selmer group of E relative to p^∞) is defined by **Definition 1.5.**

$$\mathcal{S}_{E,p} = \mathrm{Ker}\left(H^1(G_\mathbb{Q}, E_{p^\infty}) \to \prod_q H^1(G_{\mathbb{Q}_q}, E(\overline{\mathbb{Q}}_q)) \right)$$

$$G_\mathbb{Q} = \mathrm{Gal}(\overline{\mathbb{Q}}/\mathbb{Q}), \quad G_{\mathbb{Q}_q} = \mathrm{Gal}(\overline{\mathbb{Q}}_q/\mathbb{Q}_q) .$$

One sees immediately that

$$\mathrm{Coker}(i_{E,p}) = \text{Ш}(E)(p) = p\text{-primary subgroup of } \text{Ш}(E)$$

where $\text{Ш}(E)$ is the *Tate-Shafarevich group* of $E/\mathbb{Q}$, defined by

$$\text{Ш}(E) = \mathrm{Ker}\left(H^1(G_\mathbb{Q}, E(\overline{\mathbb{Q}})) \to \prod_q H^1(G_{\mathbb{Q}_q}, E(\overline{\mathbb{Q}}_q)) \right) .$$

Classical arguments in algebraic number theory show that

$$\mathcal{S}_{E,p} = (\mathbb{Q}_p/\mathbb{Z}_p)^{s_{E,p}} \oplus \text{ finite group} ,$$

for some finite integer $s_{E,p} \geq 0$. Clearly

$$s_{E,p} = g_E + t_{E,p}$$

where

$$\text{Ш}(E)(p) = (\mathbb{Q}_p/\mathbb{Z}_p)^{t_{E,p}} \oplus \text{ finite group} .$$

Thus an upper bound for $s_{E,p}$ gives an upper bound for g_E, and if we are lucky enough to find a prime p for which we can prove $t_{E,p} = 0$ then we have calculated g_E. This is exactly how classical numerical calculations of g_E proceed, taking $p = 2, 3$.

Conjecture 1.6. $t_{E,p} = 0$ *for every p.*

In fact, it is even conjectured that $\mathrm{III}(E)$ is finite, and this is one of the fundamental problems of number theory.

We still know remarkably little about $s_{E,p}$. One beautiful recent result of the Dokchitser brothers shows that the *parity* of $s_{E,p}$ is independent of p, and can be simply determined by E. For example, for the curves (1.1) with $N \equiv 5, 6, 7 \mod 8$, $s_{E,p}$ is always *odd*.

In principle, classical Galois cohomology gives an upper bound for $s_{E,p}$ for every p, but it is so bad and difficult to estimate that no one has ever written down such a bound.

2. *L-functions*

What allows us to say much more *conjecturally*, and to make some modest progress on these conjectures, is the connection of these problems with *L-functions*. We owe this great discovery to the Cambridge mathematicians B. Birch and P. Swinnerton-Dyer.

Let N_E be the conductor of E (N_E is divisible precisely by the primes where E has bad reduction). The complex L-function $L(E, s)$ is defined in the half plane $R(s) > \frac{3}{2}$ by the Euler product

$$L(E, s) = \prod_{q \mid N_E} (1 - \varepsilon_q q^{-s})^{-1} \prod_{(q, N_E)=1} (1 - a_q q^{-s} + q^{1-2s})^{-1},$$

where $\varepsilon_q = 0, \pm 1$ according as E has additive, split or non-split multiplicative reduction at the bad prime q, and, for a good prime q, a_q is defined by

$$N_q = q + 1 - a_q,$$

where N_q is the number of $\mathbb{F}_q$-rational points on the reduction of E mod q. By the deep theorem of Wiles *et al.* [1], E is modular and hence $L(E, s)$ is entire and satisfies the following functional equation. Put

$$\Lambda(E, s) = (2\pi)^{-s} \Gamma(s) L(E, s).$$

Theorem 2.1 (Wiles *et al.*). $\Lambda(E, s) = \omega_E N_E^{1-s} \Lambda(E, 2 - s)$, *where* $\omega_E = \pm 1$.

We define

Definition 2.2. $r_{e,\infty} = \mathrm{ord}_{s=1}(L(E, s))$.

Since $s = 1$ is in the centre of the critical strip, the functional equation shows that

$$\omega_E = (-1)^{r_{E,\infty}}.$$

Conjecture 2.3 (Birch and Swinnerton-Dyer). $r_{E,\infty} = g_E$.

They also conjectured an *exact formula* for the order of $\text{III}(E)$ in terms of the leading coefficient of the expansion of $L(E, s)$ about $s = 1$. *Numerically*, their formula shows that $\text{III}(E) = 0$ for most E, especially if $g_E \geq 2$. For example, it predicts that $\text{III}(E) = 0$ for the curve

$$E : y^2 = x^3 + 14x$$

with $g_E = 2$.

We are *far* from proving the conjecture of Birch and Swinnerton-Dyer. However, a few important results in its direction have been proven.

Theorem 2.4 (T. Dokchitser, V. Dokchitser [2]). *We have*

$$r_{E,\infty} \equiv s_{E,p} \quad \text{mod } 2$$

for all primes p.

Theorem 2.5 (Gross-Zagier, Kolyvagin [4,5]). *If* $r_{E,\infty} \leq 1$, *then* $g_E = r_{E,\infty}$ *and* $\text{III}(E)$ *is finite.*

Corollary 2.6. *If* $g_E \geq 2$, *then* $r_{E,\infty} \geq 2$.

Let me stress what is *unknown* by citing two specific questions:

(i) $\text{III}(E)$ has not been proven to be finite for a single E with $g_E \geq 2$.
(ii) No general inequality of the form

$$g_E \leq r_{E,\infty}$$

has ever been proven.

Note that the analogue of this last inequality was proven in the geometric case over 40 years ago by Artin and Tate [3].

The essential difficulty in attacking the conjecture of Birch and Swinnerton-Dyer is that we know no general way of relating the arithmetic objects g_E and $\text{III}(E)$ to the behaviour of the complex L-function $L(E, s)$ at $s = 1$. When some modest connexion has been established it is always by the most indirect and tortuous arguments (which are often beautiful in their own right, nevertheless).

3. Iwasawa theory

The whole theory started with Iwasawa's formulation and partial proof (the first complete proof was given by B. Mazur and A. Wiles [6] about 15 years later) of his "main conjecture" on cyclotomic fields. Let p be any prime number, and let μ_{p^∞} denote the Galois module of all p-power roots of unity. Iwasawa's main conjecture asserted a precise relation between a purely arithmetic module attached to the field $\mathbb{Q}(\mu_{p^\infty})$ and the Kummer-Leopold-Kubota *p-adic analogue* of the Riemann zeta function $\zeta(s)$. It soon became apparent that Iwasawa's revolutionary ideas could be applied to elliptic curves E, with the first important steps in this direction being made by Mazur [7]. Since then there has been a large and on going body of research investigating what turns out to be a vast and rich variety of questions.

One reason why the p-adic theory is so vast is that it seems that there is a "main conjecture" for E not only over the field $\mathbb{Q}(\mu_{p^\infty})$, but also over any infinite Galois extension of $\mathbb{Q}$ whose Galois group is a p-adic Lie group (see [13]). One can sometimes exploit these other extensions to prove deep statements about the arithmetic of E over $\mathbb{Q}$. All we have time to do in the remainder of this lecture is to state some results and conjectures which emerge from this work.

Firstly, deep work of Kato [8], combined with an important analytic theorem of Rohrlich [9], enables one to prove a striking generalization of Mordell's theorem.

Theorem 3.1 (Kato, Rohrlich). *For every prime p, $E(\mathbb{Q}(\mu_{p^\infty}))$ is a finitely generated abelian group.*

Let p be a prime with $(p, N_E) = 1$. We say that E has good ordinary reduction at p if a_p is a p-adic unit. It seems that the Iwasawa theory for E is more easily studied for good ordinary primes p.

Here is an example. Recall that we have $s_{E,p} = g_E + t_{E,p}$ (conjecturally $t_{E,p} = 0$) and we would like to have a good upper bound for $s_{E,p}$ as p-varies. We say E has *complex multiplication* if $\text{End}_{\mathbb{Q}}(E) \neq \mathbb{Z}$ and then $K = \text{End}_{\overline{\mathbb{Q}}}(E) \otimes_{\mathbb{Z}} \mathbb{Q}$ is an imaginary quadratic field. Recently, we realized that an old method of Wiles [10] and myself gave the following result:

Theorem 3.2 (Coates, Liang, Sujatha [11]). *Assume that E has complex multiplication. Then for all sufficiently large good ordinary primes p, we have*

$$s_{E,p} < 2p.$$

We do not know how to prove this result for the supersingular primes p for E, nor for E without complex multiplication. The principle behind the proof of this theorem typifies what we hope should always be true for Iwasawa theory. However, we continue to assume E has complex multiplication.

Assuming p ordinary for E, we can construct a p-adic analogue of $L(E, s)$, which we denote by $L_p(E, s)$ (s is a variable in $\mathbb{Z}_p$ in this latter function).

Definition 3.3. $r_{E,p} = \mathrm{ord}_{s=1}L_p(E, s)$.

One can then show by a simple argument that, for all sufficiently large good ordinary p, we have

$$r_{E,p} < (1 + \varepsilon)p \quad \text{for each} \quad \varepsilon > 0. \tag{3.1}$$

On the other hand, a main conjecture of Iwasawa theory (stated and partly proved by Wiles and myself, and completely proven by Rubin [14]) gives a precise connexion between $L_p(E, s)$ and the Selmer group of E over a certain subfield of $K(E_{p^\infty})$. From this it follows immediately that

$$s_{E,p} \le r_{E,p} . \tag{3.2}$$

Combining (3.1) and (3.2), we obtain the theorem.

Another feature of Iwasawa theory is that it often enables us to do *numerical* calculation which are quite impossible by classical methods. For example, if E has complex multiplication and p is a good ordinary prime, then if we can check numerically that $r_{E,p} = g_E$ for a given p, it follows from (3.2) that $\mathrm{III}(E/\mathbb{Q})(p)$ is finite. Moreover, a general formula of Perrin-Riou then gives us an exact formula for the order of $\mathrm{III}(E/\mathbb{Q})(p)$ in terms of the leading coefficient of the expansion of $L_p(E, s)$ about $s = 1$.

As an example of this, one can prove:

Theorem 3.4 (Coates, Liang, Sujatha [11]). *For the curve* $E : y^2 = x^3 + 14x$ *with* $g_E = 2$, *we have* $\mathrm{III}(E)(p) = 0$ *for all primes p with* $p \equiv 1 \mod 4$ *and* $p < 13{,}500$.

Of course, as mentioned earlier, the conjecture of Birch and Swinnerton-Dyer predicts that $\mathrm{III}(E) = 0$ for this curve.

It may be interesting to note that our computations show that for this curve, and the range of primes considered in the theorem, $L_p(E, s)$ has no other zero for s in $\mathbb{Z}_p$ other that the zero of order 2 at $s = 1$, except for the primes $p = 29{,}277$ (where there are additional zeroes).

Another surprising phenomena which arises from Iwasawa theory is that it can sometimes predict the exact growth of $E(L)$ when L runs over the finite extension of some infinite tower of fields over $\mathbb{Q}$ quite different from the tower $\mathbb{Q}(\mu_{p^\infty})$ (see [12]). We illustrate this with Leonardo Pisano's elliptic curve

$$E : y^2 = x^3 - x. \tag{3.3}$$

It was Fermat's proof that $g_E = 0$ for this curve which started our whole subject (and also led Fermat to his "last theorem"). If L is any finite extension of $\mathbb{Q}$, the Selmer group $\mathcal{S}_{E/L,p}$ is defined in the obvious fashion, and we write $s_{E/L,p}$ for its $\mathbb{Z}_p$-corank.

Theorem 3.5. *Let E be the curve (3.3). Let D be a 5-power free integer > 1, all of whose prime factors q satisfy either $q = 5$ or $q \equiv 11 \mod 20$. Put $L_n = \mathbb{Q}(\sqrt[5^n]{D})$. Then, for all integer $n \geq 1$, we have*

$$s_{E/L_n,5} = n.$$

Of course, we should have that $E(L_n)$ has rank exactly n for all $n \geq 1$, but this is unknown (we do not even know how to prove that $E(L_1)$ has rank 1). We also do not know how to prove that the complex L-function of E over L_n has a zero of exact order n at $s = 1$ (although we know there is a zero of order at least n at $s = 1$).

4. A conjectural generalization of Rohrlich's theorem

So far we have largely discussed how the complex L-functions suggest properties of the p-adic Iwasawa theory. We now want to discuss a general phenomena going in the other direction.

Suppose we are given an Artin representation (*i.e.* one which factors through a finite quotient of $G_\mathbb{Q}$)

$$\rho : G_\mathbb{Q} \to \operatorname{Aut}(W_\rho)$$

where W_ρ is a finite dimensional vector space over $\overline{\mathbb{Q}}_p$. We define the complex L-function

$$L(E, \rho, s)$$

to be the Euler product attached to the compatible system of p-adic representations $\{H_p^1(E) \otimes_{\mathbb{Q}_p} W_\rho\}$. It is conjectured to be entire and satisfy a standard functional equation, but this is so far only proven to hold in a few cases. We then define:

Definition 4.1. $r_{E,\rho,\infty} = \operatorname{ord}_{s=1} L(E, \rho, s)$.

Let us now assume ρ is irreducible, and let F be any finite Galois extension of $\mathbb{Q}$ such that ρ factors through Gal $(F/\mathbb{Q})$. Let $X_p(E/F)$ be the Pontrjagin dual of the Selmer group $\mathcal{S}_{E/F,p}$. We then define:

Definition 4.2. $s_{E,\rho,p}$ = number of copies of W_ρ in $X_p(E/F) \otimes_{\mathbb{Z}_p} \overline{\mathbb{Q}}_p$.

We then have a generalized Birch-Swinnerton-Dyer conjecture:

Conjecture 4.3. $r_{E,\rho,\infty} = s_{E,\rho,p}$ *for every prime p.*

Now let $\mathbb{Q}^{\text{cyc}}$ denote the cyclotomic $\mathbb{Z}_p$-extension of $\mathbb{Q}$, and let $\mathcal{X}$ denote the group of all 1-dimensional Artin characters of $\Gamma = $ Gal $(\mathbb{Q}^{\text{cyc}}/\mathbb{Q})$.

Theorem 4.4 (Rohrlich [9]). *For every prime p, we have $\sum_{x\in\mathcal{X}} r_{E,x,\infty}$ is finite.*

We end this lecture by conjecturing a vast generalization of this theorem. Let F_∞ be any Galois extension of $\mathbb{Q}$ such that (i) $G = $ Gal $(F_\infty/\mathbb{Q})$ is a p-adic Lie group, (ii) F_∞ contains $\mathbb{Q}^{\text{cyc}}$, and (iii) F_∞ is unramified outside a finite set of primes of $\mathbb{Q}$. We define $A(G)$ to be the set of all irreducible Artin representations of G. In view of our assumption (i), we have $\mathcal{X} \subset A(G)$.

Conjecture 4.5. *There exists a constant $C_p(E/F_\infty)$, depending only on p, E and F_∞ such that, for all ρ in $A(G)$, we have*

$$\sum_{x\in\mathcal{X}} r_{E,\rho x,\infty} \le C_p(E/F_\infty).$$

In view of the generalized conjecture of Birch and Swinnenton-Dyer, it is also natural to conjecture that we have

$$\sum_{x\in\mathcal{X}} s_{E,\rho x,p} \le C_p(E/F_\infty).$$

In the special case when $F_\infty = \mathbb{Q}(\mu_{p^\infty}, \sqrt[p^\infty]{m})$, the extension of $\mathbb{Q}$ obtained by adjoining μ_{p^∞} and all p-power roots of some fixed integer $m > 1$, this last assertion is proven in some cases in the paper [12]. The value of $C_p(E/F_\infty)$ in these cases is shown to be closely related to a simple invariant of the Iwasawa theory of E/F_∞, which can easily be calculated explicitly.

References

As the literature on the subject is vast, we only give with apologies a few recent references.

[1] A. WILES, *Modular elliptic curves and Fermat's Last Theorem*, Ann. of Math. **141** (1995), 443–551.

[2] T. DOKCHITSER and V. DOKCHITSER, *On the Birch-Swinnerton-Dyer quotients modulo squares*, Ann. of Math, to appear.

[3] J. TATE, *On the conjecture of Birch and Swinnerton-Dyer and geometric analogy*, Séminaire Bourbaki 306 (1966).

[4] B. GROSS and D. ZAGIER, *Heegner points and derivatives of L-series*, Invent Math. **84** (1986), 225–320.

[5] V. KOLYVAGIN, *Euler systems*, In: "Grothendieck Festschrift", Vol. II, Progress in Math. 87 (1990), 435–483.

[6] B. MAZUR and A. WILES, *Class fields of abelian extensions of* $\mathbb{Q}$, Invent. Math. **76** (1984), 179–330.

[7] B. MAZUR, *Rational points of abelian varieties with values in towers of number fields*, Invent. Math. **18** (1972), 183–266.

[8] K. KATO, *p-adic Hodge theory and the values of zeta functions of modular forms*, Astérisque **295** (2004), 117–290.

[9] D. ROHRLICH, *On L-functions of elliptic curves and cyclotomic towers*, Invent. Math. **75** (1984), 409–423.

[10] J. COATES and A. WILES, *On the conjecture of Birch and Swinnerton-Dyer*, Invent. Math. **39** (1977), 223–251.

[11] J. COATES, Z. LIANG and R. SUJATHA, *The Tate-Shafarevich group for elliptic curves with complex multiplication*, to appear.

[12] J. COATES, T. FUKAYA, K. KATO and R. SUJATHA, *Root numbers, Selmer groups, and non-commutative Iwasawa theory*, Journal Alg. Geometry, to appear.

[13] J. COATES, T. FUKAYA, K. KATO, R. SUJATHA and O. VENJAKOB, *On the* GL_2 *main conjecture for elliptic curves without complex multiplication*, Pub. Math. IHES **101** (2005), 163–208.

[14] K. RUBIN, *The "main conjectures" of Iwasawa theory for imaginary quadratic fields*, Invent. Math. **103** (1991), 25–68.

Colloquium
De Giorgi

2008

Partition functions, polytopes and box–splines

Claudio Procesi

1. Introduction

This paper is a short description of a body of material which will appear in a forthcoming book joint with C. De Concini [11]. The central problem we are going to discuss is that of the computation of the number of integral points in suitable families of variable polytopes. This problem is formulated in terms of the study of a suitable partition function. In a sequence of papers Dahmen and Micchelli arrive to a proof of the quasi–polynomial nature of partition functions [15].

This theorem was rediscovered and proved in a completely different way by various authors, unaware of the work of Dahmen and Micchelli, a few years later. The approach I prefer is taken from a joint paper with M. Vergne [13].

2. Partition functions and applications

2.1. Partition functions

The partition function $\mathcal{T}_X(b)$, associated to a finite set of integral vectors X, counts the number of ways in which a variable vector b can be written as linear combination of the elements in X with positive integer coefficients. Since we want this number to be finite we assume that the vectors X generate a *pointed cone* $C(X)$.

Definition 2.1. The *partition function* on $\mathbb{Z}^s$, is given by:

$$\mathcal{T}_X(b) := \# \left\{ (n_1, \ldots, n_m) \mid \sum n_i a_i = b, \ n_i \in \mathbb{N} \right\}. \tag{1}$$

The theory then consists in finding regions where the partition function is a quasi–polynomial (that is a polynomial on the cosets of a sublattice of finite index) and giving explicit formulas.

The author is partially supported by the Cofin 40 %, MIUR.

A quasi–polynomial is a function on Γ which coincides with a polynomial on each coset of some sublattice of finite index in Γ.

The main theorem [13, 27, 33] generalizing the theory of the Ehrhart polynomials [23], shows that $\mathcal{P}_X(\gamma)$ is a quasi–polynomial on the regions $\mathfrak{c} - B(X)$, where $B(X) := \{\sum_{i=1}^{m} t_i a_i, \ 0 \leq t_i \leq 1\}$ is the *zonotope* generated by X while $\mathfrak{c}$ denotes a *big cell*, that is a connected component of the complement in V of the *singular vectors* which are formed by the union of all cones $C(Y)$ for all the sublists Y of X which do not span V.

The complement of $C(X)$ is a big cell. The other cells are inside $C(X)$ and are convex.

2.2. Difference equations

The quasi–polynomials describing the partition function belong to a remarkable finite dimensional space introduced and described by Dahmen–Micchelli [27] and which in this paper will be denoted by $DM(X)$. This is the space of solutions of a system of difference equations. In order to describe it, let us call a subspace $\underline{r}$ of V *rational* if $\underline{r}$ is the span of a sublist of X. We need to recall that a *cocircuit* Y in X is a sublist of X such that $X \setminus Y$ does not span V and Y is minimal with this property. Thus Y is of the form $Y = X \setminus H$ where H is a rational hyperplane. Given $a \in \Gamma$, the *difference operator* ∇_a is the operator on functions defined by $\nabla_a(f)(b) := f(b) - f(b - a)$. For a list Y of vectors, we set $\nabla_Y := \prod_{a \in Y} \nabla_a$. We then define:

$$DM(X) := \{f \mid \nabla_Y f = 0, \text{ for every cocircuit } Y \text{ in } X\}.$$

It is easy to see that $DM(X)$ is finite dimensional and consists of quasi–polynomial functions (*cf.* [19]). All these results can be viewed as generalizations of the Theory of the Ehrhart polynomial [22, 23].

In their work Dahmen and Micchelli treat partition functions with methods that generalize their approach to two special classes of functions which play an important role in approximation theory: the *multivariate–spline* $T_X(x)$ and the *box–spline* $B_X(x)$ (introduced by de Boor, deVore [16]) associated to the given set of vectors X.

The *multivariate spline* (*cf.* [17]) is characterized by the formula:

$$\int_{\mathbb{R}^s} f(x) T_X(x) dx = \int_0^\infty \dots \int_0^\infty f\left(\sum_{i=1}^{m} t_i a_i\right) dt_1 \dots dt_m, \qquad (2)$$

where $f(x)$ varies in a suitable set of test functions (usually continuous functions with compact support or more generally, exponentially decreasing on the cone $C(X)$).

The *box spline* $B_X(x)$, is implicitly defined by the formula:

$$\int_{\mathbb{R}^s} f(x)B_X(x)dx := \int_0^1 \cdots \int_0^1 f\left(\sum_{i=1}^m t_i a_i\right) dt_1 \ldots dt_m, \qquad (3)$$

where $f(x)$ varies in a suitable set of test functions.

One of the goals of the theory is to give computable closed formulas for all these functions and, at the same time, describe some of their qualitative behavior and applications.

The main result of the theory is that, these three functions can be described in a combinatorial way as a finite sum over *local pieces* (see formulas (4) and (5)).

$$T_X(x) = \sum_{\underline{b}\ |\ \Omega \subset C(\underline{b})} |\det(\underline{b})|^{-1} p_{\underline{b},X}(-x), \qquad (4)$$

where the sum is on a combinatorial object extracted from X that is unbroken bases $\underline{b}$. $p_{\underline{b},X}(x)$ is a homogeneous polynomial of degree $|X| - s$ uniquely defined but suitable formulas. A similar formula for partition functions:

$$\mathcal{T}_X(x) = \sum_{\phi \in \tilde{P}(X)} e^{\langle \phi\ |\ x\rangle} \sum_{\underline{b} \in \mathcal{N}\mathcal{B}X_\phi\ |\ \Omega \subset C(\underline{b})} |\det(\underline{b})|^{-1} \mathfrak{q}_{\underline{b},\phi}(-x). \qquad (5)$$

In the case of $B_X(x)$ and $T_X(x)$ the local pieces span, together with their derivatives, a finite dimensional space $D(X)$ of polynomials. In the case of $\mathcal{T}_X(v)$ they span, together with their translates, the finite dimensional space $DM(X)$.

One of the important theorems characterizes:

- $D(X)$ by differential equations.
- $DM(X)$ by difference equations.

In particular Dahmen and Micchelli compute the dimension of both spaces. This dimension has a simple combinatorial interpretation in terms of X. They also decompose $DM(X)$ as a direct sum of natural spaces associated to certain special points $P(X)$ in the torus whose character group is the lattice spanned by X. In this way $DM(X)$ can be identified to a space of distributions supported at these points. $D(X)$ is then the space supported at the identity. The papers of Dahmen and Micchelli are a development of the theory of splines, initiated by Schoenberg I. J. [32]. There is a rather large literature on these topics by several authors such as A. A. Akopyan; Ben-Artzi, Asher; C. K. Chui, C. De Boor, H. Diamond,

N. Dyn, K. Höllig, Jia, Rong Qing, A. Ron, A. A. Saakyan. The interested reader can find a lot of useful historical information about these matters in the book [17] (and also the notes of Ron [31]).

The results about the spaces $D(X)$ and $DM(X)$ which, as we have mentioned originate from the theory of splines, turn out to have some interest also in the theory of hyperplane arrangements and in commutative algebra in connection with the study of certain Reisner–Stanley algebras [12] associated to hyperplane arrangements.

2.3. Connection with the index theorem

In fact the space $DM(X)$ has an interpretation in the theory of the index of transversally elliptic operators (see [13]).

In 1968 appears the fundamental work of Atiyah and Singer on the index theorem of elliptic operators, a theorem formulated in successive steps of generality [1, 2]. One general and useful setting is for operators on a manifold M which satisfy a symmetry with respect to a compact Lie group G and are elliptic in directions transverse to the G-orbits. The values of the index are generalized functions on G.

In his Lecture Notes [26] describing joint work with I. M. Singer, Atiyah explains how to reduce general computations to the case in which G is a torus, and the manifold M is a complex linear representation $M_X = \oplus_{a \in X} L_a$, where $X \subset \hat{G}$ is a finite list of characters and L_a the one dimensional complex line where G acts by the character $a \in X$. He then computes explicitly in several cases and ends his introduction saying

" ... for a circle (with any action) the results are also quite explicit. However for the general case we give only a reduction process and one might hope for something explicit. This probably requires the development of an appropriate algebraic machinery, involving cohomology but going beyond it."

It turns out that the space $DM(X)$ and some suitable generalization of it furnishes a complete answer [18].

These relations bring us to another side of our story, that having to do with the theory of hyperplane arrangements and their relations with partition functions. The fact that such relations should exist is pretty clear once we consider the set of vectors X as a set of linear equations which hence define an arrangement of hyperplanes in the space dual to that in which X lies.

In this respect we have been greatly inspired by the results of Orlik–Solomon on cohomology [28, 29] and those of Brion, Szenes, Vergne on partition functions.

In fact a lot of work in this direction originated from the seminal paper of Khovanskiĭ, Pukhlikov [30], interpreting the counting formulas for partition functions as Riemann–Roch formulas for toric varieties, and of Jeffrey–Kirwan [25] and Witten [35], on moment maps. For these matters refer to Vergne's survey article [34].

2.4. A summary

Due to the somewhat large distance between these fields, people working in hyperplane arrangements or in index theory do not seem to be fully aware of the results on the Box–spline and its combinatorial developments.

On the other hand there are some methods which have been developed in this latter theory which we believe shed some light on the space of functions used to build the box spline.

Here is a rough description of the method we follow to compute the functions which are the object of study.

- We interpret all the functions as *tempered distributions supported in the pointed cone* $C(X)$.
- We apply Laplace transform and change the problem to one in algebra, essentially a problem of developing certain special rational functions into partial fractions.
- We solve the algebraic problems by module theory under the algebra of differential operators or of difference operators.
- We interpret the results inverting Laplace transform directly.

There is a class of examples which is particularly interesting. These are *root systems* see for example [10] or [24].

Finally let us point out the recent book by Matthias Beck and Sinai Robins, in the Springer Undergraduate Texts in Mathematics: *Computing the Continuous Discretely. Integer-Point Enumeration in Polyhedra* [9].

The actual computation of partition functions or of the number of points in a given polytope is a very difficult question due to the high complexity of the problem. A significant contribution is found in the work of Barvinok [4–8] and in the papers of Welleda Baldoni [3], Charles Cochet [14], De Loera [21], Vergne M.

A useful survey paper is the one by Jésus A. De Loera, *The Many Aspects of Counting Lattice Points in Polytopes*, Mathematische Semesterberichte manuscript [20].

References

[1] M. F. SINGER and I. M. ATIYAH *The index of elliptic operators*, Ann. of Math. **87** (1968), 484–530.

[2] M. F. SINGER and I. M. ATIYAH, *The index of elliptic operators iii*, Ann. of Math. **87** (1968), 546–604.

[3] W. BALDONI-SILVA, J. A. DE LOERA and M. VERGNE, *Counting integer flows in networks*, Found. Comput. Math. **4** (2004), 277–314.

[4] A. I. BARVINOK, A. M. VERSHIK and N. E. MNËV., *Topology of configuration spaces, of convex polyhedra and of representations of lattices*, Trudy Mat. Inst. Steklov. **193** (1992), 37–41.

[5] A. BARVINOK, *Computing the Ehrhart quasi-polynomial of a rational simplex*, Math. Comp. **75** (2006), 1449–1466 (electronic).

[6] A. BARVINOK, M. BECK, C. HAASE, B. REZNICK and V. WELKER (eds.), "Integer Points in Polyhedra–Geometry, Number Theory, Algebra, Optimization", Vol. 374, Contemporary Mathematics, Providence, RI, 2005. American Mathematical Society.

[7] A. BARVINOK and J. E. POMMERSHEIM, *An algorithmic theory of lattice points in polyhedra*, In: "New Perspectives in Algebraic Combinatorics" (Berkeley, CA, 1996–97), Vol. 38 of *Math. Sci. Res. Inst. Publ.*, pages 91–147. Cambridge Univ. Press, Cambridge, 1999.

[8] A. I. BARVINOK, *A polynomial time algorithm for counting integral points in polyhedra when the dimension is fixed*, Math. Oper. Res. (4) **19** (1994), 769–779.

[9] M. BECK and S. ROBINS, "Computing the Continuous Discretely", Undergraduate Texts in Mathematics, Springer, New York, 2007. Integer-point enumeration in polyhedra.

[10] N. BOURBAKI, *Éléments de mathématique*, Fasc. XXXIV, *Groupes et algèbres de Lie*, Chapitre IV: *Groupes de Coxeter et systèmes de Tits*, Chapitre V: *Groupes engendrés par des réflexions*, Chapitre VI: *Systèmes de racines*, Actualités Scientifiques et Industrielles, No. 1337. Hermann, Paris, 1968.

[11] C. DE CONCINI and C. PROCESI, *Topics in hyperplane arrangements, polytopes and box–splines*, Forthcoming book (http://www.mat.uniroma1.it/ procesi/dida.html).

[12] C. DE CONCINI and C. PROCESI, *Hyperplane arrangements and box–splines*, Michigan Math. J. (2008), to appear.

[13] C. DE CONCINI, C. PROCESI and M. VERGNE, *Vector partition function and generalized Dahmen-Micchelli spaces*, arXiv:0805.2907.

[14] C. COCHET, "Réduction des Graphes de Goresky–Kottwitz–MacPherson: Nombres de Kostka et Coefficients de Littlewodd-Richardson", 2003, Thèse de Doctorat: Mathématiques, Universit Paris 7, Paris, France.

[15] W. DAHMEN and C. A. MICCHELLI, *The number of solutions to linear Diophantine equations and multivariate splines*, Trans. Amer. Math. Soc. (2) **308** (1988), 509–532.

[16] C. DE BOOR and R. DEVORE, *Approximation by smooth multivariate splines*, Trans. Amer. Math. Soc. (2) **276** (1983), 775–788.

[17] C. DE BOOR, K. HÖLLIG and S. RIEMENSCHNEIDER, "Box Splines", Vol. 98 of *Applied Mathematical Sciences*, Springer-Verlag, New York, 1993.

[18] C. DE CONCINI, C. PROCESI and M. VERGNE, *Vector partition functions and index of transversally elliptic operators*, arXiv:0808.2545.

[19] C. DE CONCINI and C. PROCESI *Nested sets and Jeffrey-Kirwan residues*, In: "Geometric Methods in Algebra and Number Theory", Vol. 235 of *Progr. Math.*, 139–149, Birkhäuser Boston, Boston, MA, 2005.

[20] J. A. DE LOERA, *The many aspects of counting lattice points in polytopes*, Math. Semesterber. (2) 52 (2005), 175–195.

[21] J. A. DE LOERA, R. HEMMECKE, J. TAUZER and R. YOSHIDA, *Effective lattice point counting in rational convex polytopes*, J. Symbolic Comput. (4) **38** (2004), 1273–1302.

[22] E. EHRHART, *Sur un problème de géométrie diophantienne linéaire. I. Polyèdres et réseaux*, J. Reine Angew. Math. **226** (1967), 1–29.

[23] E. EHRHART, "Polynômes arithmétiques et méthode des polyèdres en combinatoire", International Scrics of Numerical Mathematics, Vol. 35, Birkhäuser Verlag, Basel, 1977.

[24] J. E. HUMPHREYS, "Reflection Groups and Coxeter Groups", Vol. 29 of *Cambridge Studies in Advanced Mathematics*, Cambridge University Press, Cambridge, 1990.

[25] L. C. JEFFREY and F. C. KIRWAN, *Localization for nonabelian group actions*, Topology (2) **34** (1995), 291–327.

[26] M. ATIYAH, "Elliptic Operators and Compact Groups", Springer L.N.M., n. 401, Springer Verlag, Basel, 1974.

[27] D. W. MASSER, *A vanishing theorem for power series*, Invent. Math. (2) **67** (1982), 275–296.

[28] P. ORLIK and L. SOLOMON, *Combinatorics and topology of complements of hyperplanes*, Invent. Math. (2) **56** (1980), 167–189.

[29] P. ORLIK and L. SOLOMON, *Coxeter arrangements*, In: "Singularities", Part 2 (Arcata, Calif., 1981), Vol. 40 of *Proc. Sympos. Pure Math.*, pages 269–291. Amer. Math. Soc., Providence, RI, 1983.

[30] A. V. PUKHLIKOV and A. G. KHOVANSKIĬ, *The Riemann-Roch theorem for integrals and sums of quasipolynomials on virtual polytopes*, Algebra i Analiz (4) **4** (1992), 188–216.

[31] A. RON, "Lecture Notes on Hyperplane Arrangements and Zonotopes", University of Wisconsin, Madison, Wi., 2007.

[32] I. J. SCHOENBERG, "Cardinal Spline Interpolation", Society for Industrial and Applied Mathematics, Philadelphia, Pa., 1973. Conference Board of the Mathematical Sciences Regional Conference Series in Applied Mathematics, No. 12.

[33] A. SZENES and M. VERGNE, *Residue formulae for vector partitions and Euler-MacLaurin sums*, Adv. in Appl. Math. (1-2) **30** (2003), 295–342. Formal power series and algebraic combinatorics (Scottsdale, AZ, 2001).

[34] M. VERGNE, *Residue formulae for Verlinde sums, and for number of integral points in convex rational polytopes*, In: "European Women in Mathematics" (Malta, 2001), 225–285, World Sci. Publ., River Edge, NJ, 2003.

[35] E. WITTEN, *Two-dimensional gauge theories revisited*, J. Geom. Phys. (4) **9** (1992), 303–368.

Recent development on boundary value problems via Kato square root estimates

Pascal Auscher

Abstract. This colloquium lecture describes the application toward wave equations that led to the formulation of the Kato conjecture on square roots of elliptic operators. We also describe the recent application of its solution to boundary value problems for elliptic equations via a new boundary operator method for a related first order system.

1. A finite dimensional model

Let us begin with some Cauchy problem for ODEs in a finite dimensional Hilbert space $\mathcal{H}$. Assume that L is a non negative self-adjoint matrix and solve the system, a model for the wave equation,

$$\begin{cases} \ddot{u} + Lu = 0, \, t \in \mathbb{R} \\ u(0) = 0, \, \dot{u}(0) = g \in \mathcal{H}. \end{cases}$$

We teach our students that one first diagonalizes L and computes u componentwise in a Hilbertian basis of eigenvectors by

$$u_j(t) = \frac{1}{2i}(e^{it\sqrt{\lambda_j}} f_j - e^{-it\sqrt{\lambda_j}} f_j)$$

with

$$f_j = \frac{1}{\sqrt{\lambda_j}} g_j.$$

All this can be formalized by introducing the functions of L and

$$u(t) = \frac{1}{2i}(e^{it\sqrt{L}} f - e^{-it\sqrt{L}} f)$$

with

$$f = \frac{1}{\sqrt{L}} g.$$

If we now let L depend smoothly on t and add a forcing term $h(t)$, the Cauchy problem becomes

$$\begin{cases} \ddot{u} + L_t u = h, t \in \mathbb{R}, \\ u(0) = 0, \dot{u}(0) = g \in \mathcal{H}. \end{cases}$$

T. Kato introduced the following variable $X = \begin{bmatrix} v \\ w \end{bmatrix} = \begin{bmatrix} \dot{u} \\ \sqrt{L_t}u \end{bmatrix}$ and,

setting $H = \begin{bmatrix} h \\ 0 \end{bmatrix}$ and $G = \begin{bmatrix} g \\ 0 \end{bmatrix}$, the system is equivalent to the first order non autonomous system

$$\begin{cases} \dot{X} = (S_t + E_t)X + H, \\ X(0) = G \end{cases}$$

with, if $\sqrt{\dot{L}_t}$ denotes the time derivative of $\sqrt{L_t}$,

$$S_t = \begin{bmatrix} 0 & -\sqrt{L_t} \\ \sqrt{L_t} & 0 \end{bmatrix} \quad \text{and} \quad E_t = \begin{bmatrix} 0 & 0 \\ 0 & \sqrt{\dot{L}_t}\sqrt{L_t}^{-1} \end{bmatrix}.$$

(See [10] where this is detailed.) The matrix S_t is skew-symmetric so classical theory for first order non autonomous systems tells us that local solvability of the Cauchy problem on some interval I follows from the boundedness of E_t, that is

$$\|\sqrt{\dot{L}_t}f\| \le C\|\sqrt{L_t}f\| \tag{1}$$

uniformly for $f \in \mathcal{H}$ and $t \in I$. If $t \mapsto L_t$ is a C^1 curve of uniformly non negative self-adjoint matrices on I, this is true. But how does the best constant C constant depend on $\dim \mathcal{H}$? This is important if we think that the Cauchy problem is obtained by some finite dimensional approxima- tion of a PDE problem for hyperbolic equations. Unfortunately, the best constant must blow up with dimension as a consequence of examples due to McIntosh (see [5]).

2. Kato's motivation

If $\mathcal{H} = L^2(\mathbb{R}^n, \mathbb{C})$, Kato was, back in the 1950's, interested in ellip- tic operators $L_t = -\operatorname{div} A_t(x)\nabla$ with $A_t(x)$ a uniformly definite posi- tive and bounded (real) matrix of x, t, where ∇ is the operator of partial derivatives of order 1 and $-\operatorname{div}$ is its adjoint. The domain of ∇ is the Sobolev space $H^1(\mathbb{R}^n)$ and the operator L_t with domain $\mathsf{D}(L_t) = \{f \in$

$H^1(\mathbb{R}^n)$; $L_t f \in L^2(\mathbb{R}^n)\}$ is self-adjoint. There is no simple description of $\mathsf{D}(L_t)$ and in particular it heavily depends on t, which makes it a useless object. However, the domain of $\sqrt{L_t}$ defined via the functional calculus (or the spectral theorem) is t-independent and equal to $H^1(\mathbb{R}^n)$, because for all $f \in \mathsf{D}(L_t)$ (which is dense in both $\mathsf{D}(\sqrt{L_t})$ and $H^1(\mathbb{R}^n)$),

$$\|\sqrt{L_t} f\|^2 = (L_t f, f) = (A_t \nabla f, \nabla f) \sim \|\nabla f\|^2 \qquad (2)$$

uniformly in t. This is what led Kato to the above formalism to solve the Cauchy problem associated to $\partial_t^2 u + L_t u = h(t)$: the solution operator would be defined on an t-independent energy space. The question whether (1) holds becomes in this setting: is it true that for all $f \in H^1(\mathbb{R}^n)$ and uniformly in t,

$$\|\sqrt{L_t} f\| \le C \|\nabla f\|? \qquad (3)$$

To prove this inequality, there is no other known method than to imbed this question into a question about operators with time independent but complex coefficients $A(x)$, assumed to be uniformly bounded, measurable and strictly accretive in the the the sense that $\mathrm{Re}(A(x)\zeta, \zeta) \ge \lambda |\zeta|^2$ for all $\zeta \in \mathbb{C}^n$ and $x \in \mathbb{R}^n$. The operator $L = -\mathrm{div}\, A(x)\nabla$ obtained is no longer self-adjoint, but is maximal accretive so that one can still define a holomorphic functional calculus and in particular its unique maximal accretive square root $\sqrt{L}$. The question becomes what is called the Kato square root conjecture: for all $f \in \mathsf{D}(L)$,

$$\|\sqrt{L} f\| \le C \|\nabla f\| \qquad (4)$$

with a constant that depends only on n, λ, $\|A\|_\infty$. If true, then back to t-dependent C^1 matrices A_t, one obtains (3), by using the analyticity of the map $A \mapsto \sqrt{L}$ and Cauchy estimates.

But now the trivial chain in (2) is no longer valid for L since L is not self-adjoint. This made the question a difficult problem in functional analysis. Its solution was obtained in one dimension by harmonic analysis methods (quadratic estimates, Carleson measures) in 1982 by Coifman, McIntosh and Meyer [8]. It is only after the introduction of $T(b)$ theorems and their generalizations that the full solution came in 2001 by the author, Hofmann, Lacey, McIntosh and Tchamitchian [4] for arbitrary dimensions: For all L as above, (4) holds and more precisely $\|\sqrt{L} f\| \sim \|\nabla f\|$. In particular, the domain of $\sqrt{L}$ is $H^1(\mathbb{R}^n)$.

Kato's formalism for solving the time dependent wave equation does indeed lead to a theorem.

3. Elliptic problems

The previous result on L also contains the L^2 boundedness and invertibility of a boundary Neumann to Dirichlet map for the following elliptic problem:

$$\begin{cases} \partial_t^2 u - Lu = 0 \text{ in } \mathbb{R}_+^{n+1} = \{(t, x) \in (0, \infty) \times \mathbb{R}^n\} \\ u(0) = f \text{ on } \mathbb{R}^n. \end{cases} \tag{5}$$

Then, a solution is given by the semigroup equation $u(x,t) = e^{-t\sqrt{L}} f(x)$ and the Neumann data equals $-\partial_t u(0, .) = \sqrt{L} f$. The Neumann to Dirichlet map alluded to is, therefore, $\nabla \sqrt{L}^{-1}$. Hence, one can solve the Neumann problem with data in $L^2(\mathbb{R}^n)$ with an L^2 estimate on $\nabla_x u$ and $\partial_t u$ with respect to t. Also this map is invertible, with inverse $-\sqrt{L}^{-1} \text{div} A$ acting on gradient fields, and one can solve the Dirichlet problem with Dirichlet data f having its gradient in $L^2(\mathbb{R}^n, \mathbb{C}^n)$ with a uniform L^2 estimate on $\nabla_x u$ and $\partial_t u$ with respect to t (this is called the regularity problem). This holds for this general class of equations with complex coefficients. Furthermore, one can obtain some control on the solution such as quadratic estimates and then uniqueness can be discussed in an appropriate class.

This elliptic problem can be embedded in a larger class of problems allowing mixed second order derivatives

$$\begin{cases} \mathcal{L}u = 0 \text{ in } \mathbb{R}_+^{n+1} \\ u(0) = f \text{ on } \mathbb{R}^n, \end{cases}$$

where

$$\mathcal{L} = -\text{div}_{t,x} \, \mathcal{B}(x)\nabla_{t,x} \equiv -\sum_{i,j=0}^{n} \partial_{x_i} \left(b_{i,j}(x) \, \partial_{x_j} \right)$$

(we use the notational convention that $t = x_0$, $x = (x_1, \ldots, x_n)$ and $\partial_t = \partial_{x_0}$) and $\mathcal{B}(x)$ stands for an $(n + 1) \times (n + 1)$ matrix of complex-valued L^∞ coefficients, defined on $\mathbb{R}^n$ (i.e., independent of the t variable), and satisfying the uniform ellipticity condition

$$\lambda |\xi|^2 \leq \text{Re}(\mathcal{B}(x)\xi \cdot \xi), \quad \|\mathcal{B}\|_{L^\infty(\mathbb{R}^n)} \leq \Lambda,$$

for some $\lambda > 0$, $\Lambda < \infty$, and for all $\xi \in \mathbb{C}^{n+1}$, $x \in \mathbb{R}^n$. The case where $\mathcal{B}$ is block diagonal (by this, we mean $b_{i,n} = b_{n,i} = 0$ if $i = 1, \ldots, n$) with $b_{0,0} = 1$ corresponds to the system (5).

In the case where $\mathcal{B}$ is real symmetric, there is a wealth of results on solvability of the Dirichet, Neumann, regularity problems by Dalhberg,

Kenig, Jerison, Pipher, Verchota,... (see for example [9]). Apparently, the methods based on harmonic measure or double layer potentials were completely disconnected from the square root problem; yet, the technology introduced in the proof of the Kato conjecture allowed for example to obtained a perturbation theory of these results for L^2 data [1], perturbing the coefficients in L^∞. This was followed by the introduction of a new boundary operator method by the author, Axelsson and Hofmann [2] and actually this method was much simplified in recent work by the author, Axelsson and McIntosh [3] thanks to an algebraic observation that we now describe.

4. The new developments

We restrict the discussion to Neumann and Dirichlet problems. For the Dirichlet problem, the approach is similar but with a different calculus. Let us come back to complex matrices B with the above specified properties. The new boundary operator method consists in writing the second order equation $\mathcal{L}u = 0$ as a first order system

$$\begin{cases} \operatorname{div}_{t,x} B(x) F(t, x) = 0 \\ \operatorname{curl}_{t,x} F(t, x) = 0 \end{cases} \tag{6}$$

by setting $F = \nabla_{t,x} u$.

In what follows, we write

$$B = \begin{bmatrix} a & b \\ c & d \end{bmatrix},$$

where a, b, c, d are L^∞ functions taking values $a(x) \in \mathcal{L}(\mathbb{C}), b(x) \in \mathcal{L}(\mathbb{C}^n, \mathbb{C}), c(x) \in \mathcal{L}(\mathbb{C}, \mathbb{C}^n), d(x) \in \mathcal{L}(\mathbb{C}^N)$ for a.a. $x \in \mathbb{R}^n$ and

$$D = \begin{bmatrix} 0 & \operatorname{div} \\ -\nabla & 0 \end{bmatrix}$$

where $\nabla = \nabla_x$ and $\operatorname{div} = -\nabla^*$ on natural maximal domains.

We note that the strict accretivity on B implies the pointwise accretivity $\operatorname{Re} a \geq \lambda$ so we may define

$$\hat{B} = \begin{bmatrix} 1 & 0 \\ c & d \end{bmatrix} \begin{bmatrix} a & b \\ 0 & 1 \end{bmatrix}^{-1}.$$

It is a remarkable (but elementary) fact that $\hat{B}(x)$ is also bounded and strictly accretive: it is equivalent to the same property for $B(x)$.

Set $G = \begin{bmatrix} a & b \\ 0 & 1 \end{bmatrix} F$. Then, because $\mathcal{B}$ is t-independent, the first order system (6) is equivalent to the evolution equation

$$\partial_t G + D\hat{\mathcal{B}}G = 0 \tag{7}$$

on $\mathbb{R}^{1+n}_+$, with the constraint that $\mathrm{curl}_x\, \mathbf{G}(t, .) = 0$, $\mathbf{G}$ being the vector consisting of the n last components of G (the tangential part of G). Examination shows that $D\hat{\mathcal{B}}$ preserves this constraint. In this representation, $\mathbf{G}(0, .) = \nabla_x u(0, .)$ while the first component (the normal part) of $G(0, .)$ is the normal data $-\partial_\nu u(0, .)$.

The operator $D\hat{\mathcal{B}}$ is a closed, bi-sectorial operator, with resolvent estimates on the Hilbert space $L^2(\mathbb{R}^n, \mathbb{C}^{n+1})$. It turns out that recent work of Axelsson, Keith, McIntosh [7] based on the solution of the Kato conjecture proves that this operator has a bounded holomorphic functional calculus. In particular if $\chi_\pm$ is the indicator function of $\{\pm\,\mathrm{Re}\,z > 0\}$ in the complex plane, then $\chi_\pm(D\hat{\mathcal{B}})$ is a bounded operator on $L^2(\mathbb{R}^n, \mathbb{C}^{n+1})$. Note that the operator $\chi_+(D\hat{\mathcal{B}}) - \chi_-(D\hat{\mathcal{B}})$ is a generalized Cauchy integral (if $\hat{\mathcal{B}} = I$, or equivalently $\mathcal{B} = I$, this is exactly the case). This means that we can solve (7) by $G(t, .) = e^{-tD\hat{\mathcal{B}}}G(0, .)$ provided $G(0, .) \in \mathcal{H}_+$, the closure of the range of $\chi_+(D\hat{\mathcal{B}})$ (which is a generalized Hardy space).

The solvability of the Neumann or regularity boundary value problem is thus equivalent to the invertibility of the operator that maps $g \in \mathcal{H}_+$ to its normal part (Neumann) or to its tangential part (regularity). This is where we need *a priori* comparison between normal and tangential parts

$$\|g_0\| \sim \|\mathbf{g}\|$$

for all $g \in \mathcal{H}_+$. Once this is obtained a continuity method yields the desired invertibility.

This comparison is called a Rellich estimate. One can prove it when $\mathcal{B}$ is constant using Fourier transform, when $\mathcal{B}$ is block diagonal using the solution of the Kato conjecture, and finally when $\mathcal{B}$ is complex hermitean (not only real symmetric) by an integration by parts argument. This allows to recover and extend solvability of real symmetric Neumann and regularity problems for L^2 data by completely different methods. Also, this method (depending analytically on the coefficients) is stable by perturbation of the coefficients. Hence this extends the perturbation results of [1] and [2] obtained at the price of involved technicalities (Still some results there are of independent interest). It would be interesting to know more cases where the Rellich estimate holds. However, Axelsson produced examples where they fail [6].

5. Conclusion

The technology developed toward the solution to the Kato square root conjecture admits a number of powerful variants, making the range of applications wider than ever. This is probably not the end of the story.

References

[1] M. ALFONSECA, P. AUSCHER, A. AXELSSON, S. HOFMANN and S. KIM, *Analyticity of layer potentials and L^2 solvability of boundary value problems for divergence form elliptic equations with complex L^∞ coefficients*, Preprint at arXiv:0705.0836v1 [math.AP].

[2] P. AUSCHER, A. AXELSSON and S. HOFMANN, *Functional calculus of Dirac operators and complex perturbations of Neumann and Dirichlet problems*, Journal of Functional Analysis **255**, 2 (2008), 374–448.

[3] P. AUSCHER, A. AXELSSON and A. MCINTOSH, *Solvability of elliptic systems with square integrable boundary data*, Preprint arXiv:0809.4968v1 [math.AP].

[4] P. AUSCHER, S. HOFMANN, M. LACEY, A. MCINTOSH and P. TCHAMITCHIAN, *The solution of the Kato square root problem for second order elliptic operators on $\mathbf{R}^n$*, Ann. of Math. (2) **156**, 2 (2002), 633–654.

[5] P. AUSCHER and P. TCHAMITCHIAN, "Square Root Problem for Divergence Operators and Related Topics", vol. 249 of *Astérisque*, Soc. Math. France, 1998.

[6] A. AXELSSON, *Non unique solutions to boundary value problems for non symmetric divergence form equations*, To appear in Transactions of the American Mathematical Society.

[7] A. AXELSSON, S. KEITH and A. MCINTOSH, *Quadratic estimates and functional calculi of perturbed Dirac operators*, Invent. Math. **163**, 3 (2006), 455–497.

[8] R. COIFMAN, A. MCINTOSH and Y. MEYER, *L'intégrale de Cauchy définit un opérateur borné sur $L^2(\mathbb{R})$ pour les courbes lipschitziennes*, Ann. Math. **116** (1982), 361–387.

[9] C. KENIG, "Harmonic Analysis Techniques for Second Order Elliptic Boundary Value Problems", Vol. 83 of CBMS - conference lecture notes, AMS, Providence RI, 1995.

[10] A. MCINTOSH, *Square roots of operators and applications to hyperbolic PDE's*, In: *Miniconference on operator theory and partial differential equations*, Center for Math. and Appl., Canberra, 1983. Australian National Univ.

Newton polygon and Jacobian problem

Shreeram S. Abhyankar

Abstract. This is an expository article giving an extended version of my talk in Pisa on 27 May 2008. After sketching the history of the Jacobian problem, I discuss the two basic tools which are employed in attacking this problem. The first is the theory of decimal and polynomial expansions culminating in the construction of approximate roots. The second is the Newton Polygon as applied by Newton to give a constructive proof of his theorem on fractional power series expansion.

0. Prologue

(0.1) Introduction. Two bivariate polynomials are said to form a Jacobian pair if their Jacobian equals a nonzero constant. They are said to form an automorphic pair if the variables can be expressed as polynomials in the given polynomials. By the chain rule we see that every automorphic pair is a Jacobian pair. The Jacobian problem asks if conversely, assuming we are in the characteristic zero case, every Jacobian pair is an automorphic pair. It turns out that a useful method for attacking this problem is to study the similarity of polynomials. Two bivariate polynomials are similar means their degree forms, *i.e.*, highest degree terms, are powers of each other when they are multiplied by suitable nonzero constants. Geometrically this amounts to saying that the corresponding plane curves have the same points at infinity counting multiplicities in a proportional manner. At any rate, the points at infinity correspond to the distinct irreducible factors of the degree form. By using Euler's theorem on homogeneous polynomials we can see the polynomials in any nonlinear Jacobian pair are similar.

(0.2) History. Before getting into technicalities, I shall first give a short history of the problem or rather the history of my acquaintance with the problem. For that we have to go back to 1965 when a German mathematician, Karl Stein who created Stein Manifolds, wrote me a letter asking a question. He said that there was an interesting 1955 paper in the Mathematische Annalen by Engel [9]. In this paper Engel claims to prove the Jacobian theorem or what is now known as the Jacobian problem or the Jacobian conjecture or whatever. Karl Stein said to me that it is an interesting theorem but he cannot understand the proof. Can I help him? He also reduced it, or generalized it, to a conjecture about complex spaces. I wrote back to Stein giving a counterexample to his complex space conjecture. But I did not look at the Engel paper. Then in 1968, Max Rosenlicht of Berkeley asked me the same question and still I did not look at the Engel paper. Finally in 1970, my own guru (= venerable teacher) Oscar Zariski asked me the same question. Then, following the precept that one must obey one's guru, I looked up the Engel paper and found it full of mistakes and gaps.

The primary mistake in the Engel paper, which was repeated in a large number of published and unpublished wrong proofs of the Jacobian problem in the last thirty-five years, is the presumed "obvious fact" that the order of the derivative of a univariate function is exactly one less than the order of the function. Being a prime characteristic person I never made this mistake. Indeed, the "fact" is correct only if the order of the function in nondivisible by the characteristic. Of course you could say that the Jacobian problem is a characteristic zero problem, and zero does not divide anybody. But zero does divide zero. So the "fact" is incorrect if the order of the function is zero, *i.e.*, if the value of the function is nonzero. Usually this mistake is well hidden inside a long argument, because you may start with a function which has a zero or pole at a given point and your calculation may lead to a function having a nonzero value at a resulting point.

A gap is a spot where you are not sure of the argument because of imprecise definitions or what have you. The gap in the Engel paper seems to be the uncritical use of the Zeuthen-Segre invariant. For this invariant of algebraic surfaces see the precious 1935 book of Zariski [14]. Over the years I have made several attempts to understand the somewhat mysterious theory of this invariant, and I still continue to do so.

In 1970-1977 I discussed the matter in my lectures in Purdue and other places. Mostly I was suggesting to the students to fix the proof and, to get them started, I proved a few small results. Notes of my lectures were taken down by Heinzer, van der Put, Sathaye, and Singh. These appeared in [1] and [2]. Then I put the matter aside for thirty years. Seeing that the problem has remained unsolved inspite of a continuous stream of wrong proofs announced practically every six months, I decided to write up my old results, together with some enhancements obtained recently, in the form of a series of three long papers [6–8], in the Journal of Algebra, dedicated to the fond memory of my good friend Walter Feit. The Colloquium De Giorgi in the Scuola Normale of Pisa has given me a welcome opportunity of introducing these papers to the young students with an invitation to further investigate the problem.

Now one of my old results says that the Jacobian conjecture is equivalent to the implication that each member of a Jacobian pair can have only one point at infinity. Another says that each member of any Jacobian pair has at most two points at infinity. Note that the first result is a funny statement; it only says that to prove the Jacobian conjecture, it suffices to show that each member of any Jacobian pair has only one point at infinity. The second result is of a more definitive nature, and it remains true even when we give weights to the variables which are different from the normal weights. Very recently I noticed that, and this is one of the enhancements, the weighted two point theorem yields a very short new proof of Jung's 1942 automorphism theorem [11]. This automorphism theorem says that every automorphism of a bivariate polynomial ring is composed of a finite number of linear automorphisms and elementary automorphisms. In a linear automorphism both variables are sent to linear expressions in them. In an elementary automorphism, one variable is unchanged and a polynomial in it is added to the second variable. In his 1972 lecture notes [13], Nagata declared the automorphism theorem to be very profound and so it did come as a pleasant surprise to me that the weighted two point theorem yields a five line proof of the automorphism theorem. For other recent enhancements let me refer to my Feit memorial papers cited above.

The present paper is only meant to whet the student's appetite.

(0.3) Basic Strategy. The one point and two points at infinity phenomena mentioned above bring to mind the fact that amongst all the conic sections, a circle, an ellipse, and a hyperbola, have two points at infinity, but a parabola has only one. This may be coupled with the fact that a circle, an ellipse, a hyperbola, all have rational parametrizations, but only the parabola has a polynomial parametrization. For instance, the hyperbola $XY = 1$ has the rational parametrization $X = Z$ and $Y = 1/Z$, while the parabola $Y^2 = X$ has the polynomial parametrization $X = Z^2$ and $Y = Z$.

Thus, looking at points at infinity, in some sense, the Jacobian problem amounts to showing that the members of any Jacobian pair are like a parabola and not like a hyperbola.

Turning to the polynomial parametrization of a parabola, let us call a plane curve C a superbola if it has a polynomial parametrization $X = P(Z)$ and $Y = Q(Z)$ where $P(Z)$ and $Q(Z)$ are polynomials of degrees $N > 0$ and $M \geq 0$ in Z with coefficients in a field K. To go from the explicit parametric equations of C to its implicit equation we take recourse of the Y-resultant $\text{Res}_Y(f, g)$ of two polynomials

$$f(Y) = a_0 Y^N + a_1 Y^{N-1} + \cdots + a_N$$
$$g(Y) = b_0 Y^M + b_1 Y^{M-1} + \cdots + b_M$$

of degrees N, M, which is defined to be the determinant of the $N + M$ by $N + M$ matrix

$$\begin{pmatrix}
a_0 & a_1 & \cdots & & a_N & 0 & \cdots & 0 \\
0 & a_0 & a_1 & \cdots & & a_N & 0 & \cdots & 0 \\
& & \cdot & & & \cdot & & \cdot \\
& & & \cdot & & & \cdot \\
& & & & \cdot \\
0 & 0 & \cdots & a_0 & a_1 & & \cdots & a_N \\
b_0 & b_1 & \cdots & b_M & 0 & & \cdots & 0 \\
0 & b_0 & b_1 & \cdots & & b_M & 0 & \cdots & 0 \\
& & \cdot & & & \cdot & & \cdot \\
& & & \cdot & & & \cdot \\
& & & & \cdot \\
0 & 0 & \cdots & b_0 & b_1 & & \cdots & b_M
\end{pmatrix}$$

formed out of the coefficients of f and g. In 1840 this was introduced by Sylvester who showed that the value of the resultant is zero iff the two polynomials have a nonconstant common factor. Referring to my new Algebra Book [5] for more properties of the resultant, we note that the implicit equation $\phi(X, Y) = 0$ of the superbola C is given by $\phi(X, Y) = (-1)^N \text{Res}_Z(P(Z) - X, Q(Z) - Y)$. To discuss the behaviour of C at infinity we put $\Phi = \Phi(X, Y) = \phi(X^{-1}, Y)$. Assuming P to be monic in Z with $K(P(Z), Q(Z)) = K(Z)$, it follows that Φ ia a monic polynomial of degree N in Y with coefficients in the meromorphic series ring $K((X))$. Referring to my user friendly Engineering Book [4] for a discussion of the fact that a superbola has only one place at infinity, which is a refinement of having only one point at infinity, it follows that Φ is irreducible in $K((X))[Y]$. Now assuming K to be algebraically closed of characteristic zero, we can apply Newton's famous theorem on fractional power series expansion to factorize Φ into linear factors.

(0.4) Newton Polygon. Around 1665, Newton first generalized his binomial theorem from integer exponents to fractional exponents and from there went on to obtain his famous theorem on fractional power series expansion. Because of his habit of not publishing his mathematics, this theorem was soon forgotten until it was rediscovered by Puiseux in 1850. Since Puiseux used Cauchy's complex integration theory, his proof was only existential and did not work for formal power series. On the other hand, since Newton's proof was based on his polygonal method, it was very constructive and applied to convergent as well as formal power series. Now the method of Newton Polygon has been resurrected and has found numerous applications.

Here is a descriptive definition of the Newton Polygon. Given any power series in two variables, we mark in the plane those integral points which occur as exponents of nonzero terms in the expansion of the power series. We call this the support of the power series and we enclose it in a half open convex polygon. The boundary of this polygon is then the original version of the Newton Polygon. When we are dealing with a two variable polynomial, rather than a power series, the support being finite, we get a closed convex polygon. When the number of variables is more than two, then the Newton Polygon gets generalized into the Newton Polyhedron.

This Polyhedron has been very useful in desingularization problems. It may be expected that, in the multivariable polynomial case, it will be very useful in studying the Jacobian Conjecture. Likewise, turning to the applied subject area of Operations Research, the Newton Polygon and the Newton Polyhedron will have very important applications to linear programming, especially in the subfield of integer programming.

Returning to the bivariate Jacobian conjecture, in 1970-1977, by heavily using the Newton Polygon, I proved the two results about points at infinity stated in (0.2) and also showed that for a Jacobian pair of degrees N, M, the Jacobian conjecture can be settled affirmatively if either $\mathrm{GCD}(N, M) \leq 8$ or $\min(N, M) \leq 52$. Moreover, by using the Newton Polygon, I proved that the Jacobian conjecture is equivalent to the implication that for any Jacobian pair of degrees N, M, the pair (N, M) is principal by which we mean that either N divides M, or M divides N. See my publications cited in (0.2) and (0.3).

Concerning Jung's automorphism theorem referred to in (0.2), let us observe that Jung proved it for zero characteristic and in 1952 it was generalized by Kulk [12] for nonzero characteristic.

(0.5) Characteristic Sequence. Inspired by the 1875-1885 work of Halphen and Smith as reported in Zariski's 1935 book [14], we introduce the newtonian charseq (= characteristic sequence) of Φ relative to $l = N$ or $-N$ as the sequence $m = (m_i)_{1 \leq i \leq h+1}$ together with its GCD sequence $d = (d_i)_{0 \leq i \leq h+2}$, its reciprocal sequence $n = (n_i)_{1 \leq i \leq h+1}$, its difference sequence $q = (q_i)_{0 \leq i \leq h+1}$, the inner product sequence $s = (s_i)_{0 \leq i \leq h+1}$

of its difference sequence, and the normalized inner product sequence $r = (r_i)_{0 \le i \le h+1}$ of its difference sequence, thus.

By Newton's theorem we can find a root $y(T)$ of $\Phi(T^N, y(T)) = 0$, i.e., $y(T)$ in $K((T))$ with $\Phi(T^N, y(T)) = 0$. By the T-support $y(T)$ we mean the set J of all integers which are the exponents of those powers of T which occur with a nonzero coefficient in the expansion of $y(T)$. By $\min J$ we denote the smallest element of J with the understanding that the min of the empty set is ∞. The T-order of $y(T)$ is defined by putting $\mathrm{ord}_T y(T) = \min J$. We put $m_0 = l$ and $m_1 = \mathrm{ord}_T y(T)$ with $d_0 = 0$ and $d_1 = N$. For $i > 1$ we inductively let $m_i = \min J_i$ where J_i is the set of all elements of J which are nondivisible by $d_i = \mathrm{GCD}(m_0, \ldots, m_{i-1})$. This gives the sequence $m_1 < m_2 < \cdots < m_h < m_{h+1} = \infty$ where h is a positive integer and m_i is an integer for $1 \le i \le h$. It also gives the sequence of positive integers $d_1 \ge d_2 > \cdots > d_{h+1} = 1$ where d_i is divisible by d_{i+1} for $1 \le i \le h$; for convenience we put $d_{h+2} = \infty$. We put $n_i = d_1/d_i$ for $1 \le i \le h+1$. We put $q_i = m_i$ for $i = 0, 1, h+1$, and $q_i = m_i - m_{i-1}$ for $2 \le i \le h$. We put $r_i = s_i = q_i$ for $i = 0, h+1$, and $s_i = \sum_{1 \le j \le i} q_j d_j$ with $r_i = s_i/d_i$ for $1 \le i \le h$.

We call h the length of all these sequences.

h is also the number of characteristic pairs of Φ, i.e., the length of the sequence $(m_i/d_{i+1}, d_i/d_{i+1})_{1 \le i \le h}$ which is called the sequence of characteristic pairs of Φ.

Note that the above sequence $n = (n_i)_{1 \le i \le h+1}$ is a sequence of positive integers with $n_1 = 1$ such that n_i divides n_{i+1} for $1 \le i \le h$. We call such a sequence a finite expansion base.

Note that for the above sequence $r = (r_i)_{0 \le i \le h+1}$ we have that r_0 is a nonzero integer and $r_1, \ldots, r_h$ are integers with $\mathrm{GCD}(r_0, \ldots, r_{i-1}) = d_i$ for $1 \le i \le h+1$ and $d_1 \ge d_2 > \cdots > d_{h+1} = 1$. Moreover, $n_i = d_1/d_{i+1}$ for $1 \le i \le h+1$, and hence n is the reciprocal sequence of r.

The situation of $l = N$ is meant for the analytic case when the coefficients of Φ belong to the power series ring $K[[X]]$.

In our present "pure meromorphic case" when the coefficients of Φ belong to the polynomial ring

$$R = K[X^{-1}]$$

we let $l = -N$.

Note that now $(r_0, r_1) = (m_0, m_1) = (-N, -M)$ and hence the pair (r_0, r_1) is principal iff the pair (N, M) is principal.

We also have: $\mathrm{GCD}(N, M) = 1 \Leftrightarrow h = 1$, and: $\mathrm{GCD}(N, M) =$ a prime number $\Rightarrow h = 2$.

(0.6) Decimal Expansion. By generalizing the usual decimal expansion of integers, we can show that every integer A has a unique n-adic expansion $A = \sum_{1 \leq i \leq h+1} z_i n_i$ where the "digits" $z_1, \ldots, z_{h+1}$ are integers with $0 \leq z_i < n_{i+1}/n_i$ for $1 \leq i \leq h$. Moreover, we have the semigroup property saying that A is nonnegative iff the "free digit" z_{h+1} is nonnegative.

The matter can be generalized by showing that every integer A has a unique r-mal expansion $A = \sum_{0 \leq i \leq h} t_i r_i$ where the "digits" $t_0, \ldots, t_h$ are integers with $0 \leq z_i < d_i/d_{i+1}$ for $1 \leq i \leq h$. But this time the semigroup property says that if A belongs to the semigroup $r\mathbb{N}$ generated by r, i.e., if $A = \sum_{0 \leq i \leq h} t_i' r_i$ for some nonnegative integers $t_0', \ldots, t_h'$, then the "free digit" t_0 is nonnegative. We call r a strict generating system in case this property holds,

It can be shown that if r is a strict generating system such that $r_i < 0$ for $2 \leq i \leq h$ and -1 belongs to $r\mathbb{N}$, then the pair (r_0, r_1) is principal. Let us call this the Principality Lemma.

(0.7) Polynomial Expansion. Let $\Psi = (\Psi_j)_{1 \leq j \leq h+1}$ where $\Psi_j = \Psi_j(X, Y)$ is a monic polynomial of Y-degree n_j with coefficients in R. By the existence and uniqueness of n-adic expansion we see that any $\Theta = \Theta(X, Y) \in R[Y]$ has a unique Ψ-adic expansion

$$\Theta = \sum_{z \in B(n)} \Theta_z \Psi^z \quad \text{with} \quad \Psi^z = \prod_{1 \leq j \leq h+1} \Psi_j^{z_j} \quad \text{and} \quad \Theta_z \in R$$

where $B(n)$ is the set of all integer sequences $z = (z_1, \ldots, z_{h+1})$ such that $z_{h+1} \geq 0$ and $0 \leq z_j < n_{j+1}/n_j$ for $1 \leq j \leq h$. Consider the integer semigroup

$$\text{Sem}_R \Phi = \{\text{int}(\Phi, \Theta) : \Theta \in R[Y] \setminus \Phi K((X))[Y]\}$$

consisting of the integers $\text{int}(\Phi, \Theta)$ as Θ varies over the set of all $\Theta \in R[Y]$ which are nondivisible by Φ in $K((X))[Y]$, where the intersection multiplicity $\text{int}(\Phi, \Theta)$ of Φ with Θ is defined by putting $\text{int}(\Phi, \Theta) = \text{ord}_X \text{Res}_Y(\Phi, \Theta)$, i.e., equivalently by putting $\text{int}(\Phi, \Theta) = \text{ord}_T \Theta(T^N, y(Y))$. Now assume that

$(\bullet)$ $\qquad\qquad \text{int}(\Phi, \Psi_j) = r_j \quad \text{for} \quad 1 \leq j \leq h+1.$

For every $z \in B(n)$ with $\Theta_z \neq 0$, let $z^* = (z_0^*, \ldots, z_h^*)$ be defined by putting $z_i^* = z_i$ or $\text{ord}_X \Theta_z$ according as $1 \leq i \leq h$ or $i = 0$, and we note then z and z^* are the n-adic and r-mal expansions of $\langle z, n \rangle = z_1 n_1 + \cdots + z_{h+1} n_{h+1}$ and $\langle z^*, r \rangle = z_0^* r_0 + \cdots + z_h^* r_h$ respectively. Consequently, if Θ

is not divisible by Φ then there is a unique $y \in B(n)$ with $\Theta_y \neq 0 = y_{h+1}$ such that for all $z \neq y$ in $B(n)$ with $\Theta_z \neq 0 = z_{h+1} = 0$ we have $\langle y^*, r \rangle < \langle z^*, r \rangle$; moreover, for this y we have $\text{int}(F, \Theta) = \langle y^*, r \rangle$. It follows that $\text{Sem}_R \Phi = r\mathbb{N}$ and

(••) r is a strict generation system with $r < 0$ for $2 \leq i \leq h$.

(0.8) Approximate Roots. The "completing the square" method of solving quadratic equations was invented by the Indian Mathematician Shreedharacharya in 500 A.D. and was versified by the Indian Mathematician Bhaskaracharya in 1150. Around 660 A.D. it was generalized by Tschirnhausen to "completing the N-th power." For any divisor D of N, by repeatedly applying the Tschirnhausen procedure to the $(Y, Y^{N/D})$-adic expansion of Φ we get the unique existence of the approximate D-th root of Φ as the monic polynomial $\text{App}_D \Phi$ of degree N/D in Y with coefficients in R such that $\deg_Y(\Phi - (\text{App}_D \Phi)^D) < N - (N/D)$. Taking $\Psi_1 = Y$ and $\Psi_j = \text{App}_{d_j} \Phi$ for $2 \leq j \leq h + 1$, as the main result in my paper [3] we get (•).

Therefore, in view of (••), by the Principality Lemma, we immediately get the HIGH SCHOOL EPIMORPHISM THEOREM which says that if Z belongs to the polynomial ring $K[P(Z), Q(Z)]$ then the pair (N, M) is principal.

As said in (0.4), the Jacobian conjecture is equivalent to the implication that for any Jacobian pair $f(X, Y), g(X, Y)$ of positive degrees N, M over a characteristic zero field k, the pair (N, M) is principal. To relate this to the above discussion, by linear transformations arrange f, g to be monic of Y-degrees N, M with coefficients in $k[X]$. Now take indeterminates W, Z, let K be an algebraic closure of $k(W)$, and take $P(Z) = f(W, Z), Q = g(W, Z)$. Now supplementing the above analysis by some more work with the Newton Polygon, we settle the Jacobian conjecture if either $h \leq 2$ (done in 1970-1977) or $h = 3$ with even d_3 (new enhancement).

The rest of the paper is mostly an exposition of [3].

References

[1] S. S. ABHYANKAR, *Some remarks on the Jacobian question,* Purdue Lecture Notes (1971), 1–20; Published in the Proceedings of the Indian Academy of Sciences **104** (1994), 515–542.

[2] S. S. ABHYANKAR, "Expansion Techniques in Algebraic Geometry", Tata Institute of Fundamental Research, Bombay, 1977.

[3] S. S. ABHYANKAR, *On the semigroup of a Meromorphic Curve* (Part I), Proceedings of the International Symposium on Algebraic Geometry (Kyoto), Kinokuniya, Tokyo, 1977, 249–414.

[4] S. S. ABHYANKAR, "Algebraic Geometry for Scientists and Engineers", American Mathematical Society, 1990.

[5] S. S. ABHYANKAR, "Lectures on Algebra I", World Scientific, 2006.

[6] S. S. ABHYANKAR, *Some Thoughts on the Jacobian Conjecture, Part I,* Journal of Algebra **319** (2008), 493–548.

[7] S. S. ABHYANKAR, *Some Thoughts on the Jacobian Conjecture, Part II,* Journal of Algebra **319** (2008), 1154–1248.

[8] S. S. ABHYANKAR, *Some Thoughts on the Jacobian Conjecture, Part III,* Journal of Algebra, to Appear.

[9] W. ENGEL, *Ein Satz über ganze Cremona Transformationes der Ebene,* Mathematische Annalen **130** (1955), 11–19.

[10] N. JACOBSON, "Basic Algebra II", Freeman 1980.

[11] H. W. E. JUNG, *Über ganze birationale Transformatione der Ebene,* Crelle Journal **184** (1942), 161–174.

[12] W. VAN DER KULK, *On polynomial rings in two variables,* Nieuw Arch. Wisk. **3** (1953), 33–41.

[13] M. NAGATA, "On the Automorphism Group of $k[X, Y]$", Lecture Notes in Mathematics, Tokyo, 1972.

[14] O. ZARISKI, "Algebraic Surfaces", Springer-Verlag, 1935.

Algebraic varieties over small fields

Fedor Bogomolov

Abstract. In this lecture I report on our joint results with Yuri Tschinkel. We discuss special geometric properties of algebraic varieties defined over "small" algebraically closed fields such as $\bar{Q}$ or $\bar{F}_p$ where p is a prime number.

Smooth complex projective algebraic variety X can be defined over a number field if it can be obtained as zero set in P^n of a family of homogeneous polynomial equations which coefficients are algebraic numbers. Such varieties are called arithmetic. There are many such varieties, but only countably many nonisomorphic complex projective varieties admit such a structure. On the contrary the family of all non-isomorphic complex projective varieties has a cardinality of continuum which coincides with the cardinality of complex numbers.

For example arithmetic algebraic curves of a given genus correspond to everywhere dense subset of algebraic points in the moduli space of such curves which in turn has a structure of arithmetic variety. Note that the field of algebraic numbers $\bar{Q}$ is a union of arithmetic subfields- finite extensions of Q. Any finite subset of elements in $\bar{Q}$ is also contained in an arithmetic subfield. Thus any curve defined over $\bar{Q}$ is in fact defined over some finite extension K of Q. In case of a complex smooth projective curve R an equivalent definition can be given in terms of the field of meromorphic fucntions on R. Namely R can be defined over K if and only if the field of complex meromorphic functions $\mathbb{C}(R)$ is equal to the tensor product $K(R) \otimes \mathbb{C}$ where $K(R)$ is a finite extension of $K(t)$ and the tensor product is taken with respect to a fixed imbedding $K \subset \mathbb{C}$. Note that different imbeddings $K \subset \mathbb{C}$ produce different (conjugated) fields of functions and different complex curves.

Belyi's theorem provided one of the first examples demonstrating a striking difference between geometry of the varieties defined over small fields and varieties with a "big" field of definition (like $Q(t_1, ..., t_i)$ or even $(F_p(t))$.

Theorem 0.1. *Let R be a complex projective curve. Then there is a map $f : R \to P^1$ with exactly three ramification points in P^1 if and only if R can be defined over a number field.*

Since any three points on $P^1(\mathbb{C})$ are projectively equivalent to a standard triple of points $(-1, 1, \infty)$ the "only if" of the theorem part follows from standard arguments of Galois theory.

The proof in the opposite direction uses a construction invented by Beliy.

Assume that R is defined over arithmetic subfield $K \subset \bar{Q}$ then the field of complex meromorphic functions $\mathbb{C}(R)$ contains a subfield $K(R)$ of functions defined over K.

Belyi proved in fact a stronger result. He showed that for any $f \in K(R) - K$ which defines a finite surjective map $f : R \to P^1$ there is a meromorphic map (function) $g : P^1 \to P^1$ such that the composition $gf : R \to P^1$ is ramified only over $(-1, 1, \infty)$.

Similar and even stronger result holds for curves R defined over $\bar{F}_p$. The field $\bar{F}_p$ is also a union of finite fields $F_{q'}$ where $q' = p^{n'}$. In this case any surjective map $f : R \to P^1$ can be supplemented by a map $g : P^1 \to P^1$ such that $gf : R \to P^1$ is ramified over two point. The proof of this statement is however much simpler since any polynomial map $h : P^1 \to P^1$ with $h = f^{p^n} + ax + d$ is ramified over exactly one point of infinity. Note that since $F_q \subset \bar{F}_p, q = p^n$ is also a finite Abelian additive group acting by translations on $P^1(\bar{F}_p)$ we have a natural map $P^1 \to P^1/F_q = P^1$ which is ramified over ∞ only and maps all points in $F_q \subset \bar{F}_p \subset P^1(\bar{F}_p)$ into one point on $P^1/F_q = P^1$.

Thus we can easily collect all other ramification points by mapping them into 0 by appropriate f and hence obtain the result.

Few years ago we obtained new results in a similar direction using a completely different method. They are also results indicating that all the curves defined over small fields with genus greater than 2 have similar arithmetic properties.

Theorem 0.2. *Any complex hyperelliptic curve H of genus $g \geq 2$ has a nonramified covering $\tilde{H}$ of degree 216 which surjects with degree 4 onto a curve C_6 of genus 2 given by affine equation $y^6 = x^2 - 1$.*

The notation above means that the field of meromorphic functions $\mathbb{C}(C_6)$ is an extension of degree 2 of $\mathbb{C}(x)$ with additional coordinate y satisfying the equation $y^6 = x^2 - 1$.

The prove is based on the following simple (and well-known) observation. (See for example [8]) Consider any complex elliptic curve E. If we fix a point zero in E then there is unique structure of commutative algebraic group with given 0. Let $\theta : x \to -x$ be a standard involution $\theta : x \to -x$ on E and $p : E \to E/\theta = P^1$ be a projection on the quotient. Then the image of the subset of 8 points of order exactly 3 in

$P^1 = E/\theta$ consists of four points which are projectively equivalent to $(1, s, s^2, \infty) \subset P^1(\mathbb{C})$ where $s^3 = 1$.

The unramified covering $\tilde{H}$ of H is constructed of compositiion of 5 Abelian coverings of order 2 and 9. The main technical trick is the following: Let $R \to C$ and $R' \to C$ be two surjective maps of curves with the same ramification points on C. Assume that for each point $p \in C$ the ramification index at any $p_i \in R$ over p is divisible by the ramification index of any $q_j \in R'$ which lies over p. Then the surjection of the fiber product $R \times_C R' \to R$ is nonramified.

In fact Belyi theorem implies the following theorem which shows that special hyperelliptic curves dominate all other curves. Let us define C_n as a hyperelliptic curve with $\mathbb{C}(C_n) = \mathbb{C}(x, y)$ with $y^n = x^2 - 1$ then $g(C_n) \geq 2$ for $n \geq 5$. The element $x \in \mathbb{C}(C_n)$ defines a projection $x : C_n \to P^1$ which is ramified over exactly three points $-1, 1, \infty$ with ramification indices n, n, n for odd n and $n, n, n/2$ for even n over points $-1, 1, \infty$ respectively.

Theorem 0.3. *For any complex projective arithmetic curve C there is an integer $n > 4$ and a finite nonramified covering $\tilde{C}_n$ of C_n with a finite surjective map onto C.*

By Belyi's theorem there is map $f : C \to P^1$ which is ramified over points $-1, 1, \infty$ only. Let n be an integer such that $n/2$ is divisble by ramification indices of f at all the branch points of f. Let us take $\tilde{C}$ be a connected component of the fiber product $C_n \times_{P^1} C$. Then $\tilde{C}$ surjects naturally on both C_n and C but by the argument above the surjection of $\tilde{C}$ to C_n is unramified.

Remark 0.4. In fact this result is true for many other classes of curves, for example for modular curves as it was first pointed by Yu. Manin in 1978. However any known class of such curves consists of an infinite number of curves. It is a challenge to find a finite set of arithmetic curves with such property as the class of curves $C_n, n > 4$ has. In particular the question is whether for any curve C_n there is a nonramified covering $\tilde{C}_{6,n}$ of C_6 such that $\tilde{C}_{6,n}$ surjects onto C_n. Partial affirmative results in this direction are contained in [3].

However such universal curves exist over $\bar{F}_q$ and in partiular the curve C_6 which is minimal in the class of hyperelliptic curves dominates any curve over $\bar{F}_q$.

Theorem 0.5. *Let $p \neq 2$ (where $q = p^n$) and C is a projective curve defined over F_q. Then for any hyperelliptic curve H of genus $g \geq 2$ there is a finite nonramified covering $\tilde{H}$ of H which surjects on C.*

The proof is based on the same trick as the proof of the previous theorem but uses the fact that all points in any elliptic curve $E(\bar{F}_q)$ have finite order as the elements of the corresponding group. (See [3, 6] for more detailed description and other related results.)

Another series of results stems from the classical works of Tate, Mumford and others on Abelian varieties defined over finite fields. They showed in particular that such Abelian varieties are of CM-type which means that their ring of endomorphisms End A is very big with End $A \times Q$ splitting into a direct sum matrix algebras over skew-fields. So that $\dim(\text{End } A \times Q) \geq 2 \dim A$.

We have used these results to show the existence of an anomally big number of rational curves on some varieties defined over the field $\bar{F}_p$.

In recent years the notion of rationally connected variety gain a lot of attention in the theory of complex projective manifolds. It means that arbitrary points on a given variety can be connected by rational curves. In the case of a complex variety it always implies triviality of the space of sections of all tensor bundles $(\Omega^1(X))^n$ where $n > 0$ is arbitrary positive integer and the power means tensor power of the bundle. The class of complex rationally connected varieties contains all unirational varieties, i.e. varieties X such that the field of meromorphic functions $C(X)$ can be imbedded into finite purely transcendental extension $C(x_1, \ldots, x_n)$ of C. Presumably the class of complex rationally connected varieties is strictly bigger but in dimensions $1, 2$ all complex rationally connected varieties are in fact rational.

This fails dramatically in the case of $\bar{F}_q$. In fact it was known that there are surfaces with many tensors over $\bar{F}_q$ (general type surfaces) which are unirational. However the proper notion of general type includes a non-triviality of some other invariants like fundamental group or infinitely divisible unramified Brauer group. Both these invariants are trivial for unirational varieties. However we produced a family of examples of rationally connected surfaces and higher dimensional varieties over $\bar{F}_q$ with both invariants above being nontrivial.

Our construction is based on the theory of Abelian varieties over the field $\bar{F}_p$.

Consider Abelian variety A defined over a finite field $F_q, q = p^n$ and an algebraic curve C generating A (in the sense that the induced map of $J(C)$ onto A is surjective) The group of points $A(\bar{F}_q)$ is torsion Abelian group with a decomposition into a sum $\Sigma_{l \neq p}(Q_l/Z_l)^{2g}$ and $(Q_p/Z_p)^r, r \leq g$ where g is dimension of A. It corresponds to a direct decomposition of $A(\bar{F}_q)$ into a sum of Sylow subgroups A_l.

Theorem 0.6 ([4,5]). *For any finite set of primes S there exists a subset of positive density R in the set primes (in fact the latter can be made of density 1 [12]) such that $S \subset R$ and the projection $C(\bar{F}_q) \to \sum l \in RA_l$ is surjective.*

The proof for one l was published by R. Indik and G. Anderson in 1985 (see [1] and also [13]) and the general result can be deduced from the twisted version of Weil-Deligne formula developed by N. Katz [10]

Corollary 0.7. *Consider a multyplication $m : A \to A$ and denote the image of C by mC. Then for any sequence $N = p(mod q)$ with p, q coprime the union $\bigcup_{mC(\bar{F}_q)}$ contains all the poinst of $A((\bar{F}_q))$.*

The result yields interesting geometric applications.

Corollary 0.8. *Assume that A is generated by a hyperelliptic curve H then the Kummer variety A/θ is covered by the union of rational curves $P_m = (mH/\theta)$ where $\theta; x \to -x$ is a standard involution on A.*

In particular this result holds for (singular) Kummer surfaces since any Abelian surface contains a generating hyperelliptic curve. In fact all the curves P_m are defined over the same finite field over which the pair A, H and the map $H \to A$ are defined. Note that the result also holds for the products of Kummer surfaces since we can pass a rational curve through an arbitrary finite subset in the singular Kummer surface. As a consquence it holds for a symmetric power of the singular Kummer $K3$-surfaces. The same argument shows that it is true also for any variety S_m which is a fiber of the surjective map $S^m A^2 \to A_2$ where A^2 is an Abelian surface. The two last series of examples are singular models over $\bar{F}_p$ of special (Kummer type) varieties of the only two known infinite series of hyperkähler manifolds.

It is also possible to show similar result for some special manifolds with trivial canonical class.

The first problem for general Kummer varieties is how to genelize the above result for a Kummer variety of arbitrary Abelian variety over $\bar{F}_p$.

Namely how to avoid the condition on the existence of a generating hyperelliptic curve in A? It amounts to the following general question:

Does any Abelian variety over $\bar{F}_p$ contain a generating hyperelliptic curve?

Pirola [11] and Oort-de Jong [9] proved that generic Abelian varieties over "big fields" don't contain hyperelliptic curves. However both proofs use arguments which work neither over $\bar{F}_p$ nor over $\bar{Q}$.

Thus it may still be true for $\bar{F}_p$ in particular since it can be reduced essentially to the following question related to the endomorphism algebra of A.

Does any semisimple algebra End $A \otimes Q$ appear as a direct summand of the algebra End $J(H) \otimes Q$ for some hyperelliptic curve H?

References

[1] G. W. ANDERSON and R. INDIK, *On primes of degree one in function fields*, Proc. Amer. Math. Soc. **94** (1985), 31–32

[2] G. V. BELYI, *Galois extensions of a maximal cyclotomic field* (Russian), Izv. Akad. Nauk SSSR Ser. Mat. **43** (1979), 267–276, 479.

[3] F. BOGOMOLOV and Y. TSCHINKEL, *Couniformization of curves over number fields*, In: "Geometric Methods in Algebra and Number Theory", 43–57, Progr. Math., Vol. 235, Birkhuser Boston, Boston, MA, 2005.

[4] F. BOGOMOLOV and Y. TSCHINKEL, *Rational curves and points on K3 surfaces*, Amer. J. Math. **127** (2005), 825–835.

[5] F. BOGOMOLOV and Y. TSCHINKEL, *Curves in Abelian varieties over finite fields*, Int. Math. Res. Not. (2005), 233–238.

[6] F. BOGOMOLOV and Y. TSCHINKEL, *On curve correspondences*, In: "Communications in arithmetic fundamental groups" (Kyoto, 1999/2001). Su-rikaisekikenkyu-sho Ko-kyu-roku No. 1267 (2002), 157–166.

[7] F. BOGOMOLOV and Y. TSCHINKEL, *Unramified correspondences*, In: "Algebraic Number Theory and Algebraic Geometry", 17–25, Contemp. Math., Vol. 300, Amer. Math. Soc., Providence, RI, 2002.

[8] D. HUSEMOLLER, "Elliptic Curves", Second edition, with appendices by Otto Forster, Ruth Lawrence and Stefan Theisen, In: "Graduate Texts in Mathematics", 111.

[9] F. OORT and J. DE JONG, *Hyperelliptic curves in Abelian varieties*, Algebraic geometry, 5, J. Math. Sci. **82** (1996).

[10] N. M. KATZ, "Twisted L-Functions and Monodromy", Annals of Mathematics Studies, Vol. 150, Princeton University Press, Princeton, NJ, 2002. viii+249 pp. ISBN: 0-691-09150-1; 0-691-09151-X

[11] G. P. PIROLA, *Curves on generic Kummer varieties*, Duke Math. J. **59** (1989), 701–708.

[12] B. POONEN, *Unramified covers of Galois covers of low genus curves*, Math. Res. Lett. **12** (2005), 475–481.

[13] F. POP and M. SAIDI, *On the specialization homomorphism of fundamental groups of curves in positive characteristic*, Galois groups and fundamental groups, 107–118, Math. Sci. Res. Inst. Publ., 41, Cambridge Univ. Press.

The Dirichlet problem for some fully nonlinear elliptic equations

Louis Nirenberg

This is a report on some results of a paper [6] of R. Harvey and B. Lawson. It concerns the Dirichlet problem for general second order nonlinear elliptic equations of the form

$$F(D^2 u) = 0.$$

Recall ellipticity for a general second order nonlinear elliptic operator of the form

$$F(x, u, Du, D^2 u).$$

Here u is a scalar function defined on a bounded domain Ω in $\mathbb{R}^n$. The operator is considered elliptic if the matrix

$$\frac{\partial F}{\partial u_{ij}}$$

is positive definite for all values of the arguments. Degenerate ellipticity means that for A and B in $S^{n \times n}$, the space of $n \times n$ symmetric matrices,

$$F(x, u, p, A) \leq F(x, u, p, A + B) \tag{1}$$

in case B is a nonnegative matrix.

Harvey and Lawson consider degenerate elliptic operators of the special form

$$F(D^2 u) = 0 \tag{2}$$

and the Dirichlet problem

$$u = \phi \quad \text{on} \quad \partial\Omega. \tag{3}$$

Before describing some results, first, a bit of background. In the 1980's – L. Caffarelli, J. Spruck, and I wrote a number of papers on fully nonlinear elliptic equations. In particular in [1] we considered equations of the form

(2) (and more general, with a given right hand side) but we considered F depending only on the (ordered) eigenvalues $\lambda : \lambda_1 \le \lambda_2 \ldots \le \lambda_n$ of the Hessian $D^2 u$,

$$\mathbb{F}(D^2 u) = f(\lambda) = 0 \tag{4}$$

f was assumed to be symmetric in the λ's under permutation, and ellipticity meant

$$f_{\lambda_i} > 0 .$$

I want to draw attention to the point of view we took: The equation (4) defines a hypersurface Σ in $\mathbb{R}^n$ - the λ-space. The function f does not matter, only Σ does; f may be replaced by a function of f. We then considered an open set G in $\mathbb{R}^n$, with boundary Σ. We assumed that the boundary of G is smooth, and that the inner normal at any point of Σ is a positive vector, *i.e.* all its components are ≥ 0. In addition we assumed that

$$G \quad \text{is convex} .$$

We obtained smooth solutions, and the convexity of G played an essential role in the derivation of a priori estimates for derivatives of the solution.

Example 1. Laplace equation: $\Sigma \lambda_i = 0$. In this case G is the half space $\Sigma \lambda_i > 0$.

Example 2. Monge-Ampère equation:

$$\text{Det}(D^2 w) = \prod \lambda_i = 1 \quad \text{say} . \tag{5}$$

Here, to have ellipticity it made sense to consider only convex functions. For the Dirichlet problem (2), in case $\phi = $ constant, $\partial\Omega$ has to be a level set of a convex function, so we considered only convex Ω.

Thus, part of the problem when trying to find u such that $\lambda(D^2 u(x))$ lies on $\partial G = \Sigma, \forall\, x \in \Omega$, is to determine for what classes of domains Ω this is possible.

The condition on Ω, – a kind of "convexity" – that we imposed was the following: At every point on the boundary $\partial\Sigma$, let $k_{y-1}, \ldots, k_{n-1}$ be the principal curvatures of $\partial\Omega$, relative to the interior normal. Then we required that for some large R,

$$(k_{y-1}, \ldots, k_{n-1}, R) \quad \text{lies in} \quad G , \tag{6}$$

– for every boundary point.

One more example:

Example 3. Let σ_j be the jth elementary symmetric function. Find a function in Ω satisfying (2), with $\lambda(D^2 u(x))$ lying on the boundary of G.

Here, G = component of points containing $(1, 1, \ldots 1)$, of vectors λ satisfying

$$\sigma_j(\lambda) > 1 .$$

In the problem, "convexity" condition (6) is equivalent to the following:

$$\sigma_{j-1}(k_1, \ldots, k_{n-1}) > 0$$

at every point on $\partial\Omega$.

The papers by Caffarelli, Spruck and myself are highly technical because of the a priori estimates. Since then, of course, much further work has been done.

In particular, B. Guan and J. S. Spruck [3] solved the Dirichlet problem (3) for Monge-Ampère equations in domains Ω which did not have to be convex. It suffices that there is a strict subsolution v (*i.e.* satisfying $\pi\lambda_i > 1$ in the problem (5)), with $v = \phi$ on $\partial\Omega$. Guan in [4] improved this result by just requiring the existence of a subsolution having the given boundary values. In [5] he treats more general classes of equations, including the σ_j described above, requiring, again, the existence of a subsolution with the given boundary values. Furthermore these papers also treat similar equations on Riemannian manifolds. Further references may be found there.

Turn now to [6]. They obtain solutions which are continuous in $\overline{\Omega}$; no further regularity is proved; *there are no a priori estimates*. The solutions are viscosity solutions.

They extend the point of view of [1].

Namely, the equation $F(D^2 u) = 0$ defines a hypersurface Σ in $S^{n \times n}$, the space of $n \times n$ symmetric matrices. Their method of obtaining solutions is one familiar in the study of viscosity solutions; the Perron method. The solution of the Dirichlet problem

$$u = \phi \quad \text{on} \quad \partial\Omega$$

is obtained as the sup of relative-to-the-problem "subharmonic" functions v which are $\leq \phi$ on $\partial\Omega$. The main tool is the maximum principle; no a priori estimates are derived.

They extend the point of view of [1] (they follow ideas of N. V. Krylov) by introducing

Dirichlet Set: This is a proper closed set $\mathcal{G}$ in $S^{n \times n}$ with the property that

$$\mathcal{G} + P \subset \mathcal{G} ,$$

where P is the set of nonnegative symmetric matrices. The set of C^2 function u in Ω with $D^2u(x) \in \mathcal{G}$, $\forall\, x$ in Ω, are called $\mathcal{G}$-subharmonic (they use the notation $\mathcal{G}$-plurisubharmonic).

A C^2 solution u should be a function with

$$D^2u(x) \in \partial\mathcal{G} \quad \forall\, x \in \Omega.$$

Related to $\mathcal{G}$-subharmonic is the nation of $\mathcal{G}$-superharmonic which they define using a

Dual Dirichlet set $\widetilde{\mathcal{G}} = -\left(\text{interior } \mathcal{G}\right)^c$. One has

$$\widetilde{\widetilde{\mathcal{G}}} = \mathcal{G}.$$

An important example is $\widetilde{P}$. As one checks easily, this consists of matrices with $\lambda_n \geq 0$. (The eigenvalue are always ordered, $\lambda_1 \leq \ldots \leq \lambda_n$.)

Now functions u with $\lambda_n(D^2u) \geq 0$ are closely related to the maximum principle and they play a basic role in the paper. They are tied to the ration of

Subaffine function. A function u in a domain Ω is subaffine if in any subdomain D, u plus any linear function has its maximum on the boundary of D.

(Note that if a function u is locally subaffine, then it is also globally. On the other hand a function satisfying the maximum principle locally need not satisfy it globally.)

Easy Remark. A C^2 function u is subaffine $\iff$ $\lambda_n(D^2u) \geq 0$ everywhere.

(Independently, Caffaarelli, Y. Y. Li and I introduced the same notion, but for other purposes, in a paper to appear.)

Next, a simple lemma, which plays however an important role.

Lemma 1. *If u and v are C^2 functions, with u, $\mathcal{G}$-subharmonic, and v; $\widetilde{\mathcal{G}}$-subharmonic then*

$$u + v \quad \text{is subaffine.}$$

(This is just a statement about matrices: A in $\mathcal{G}$, B in $\widetilde{\mathcal{G}}$; then $\lambda_n(A+B) \geq 0$.)

We now turn to the class of functions they consider. These are USC, upper semicontinuous functions.

Next we now define the notion for USC functions of

G-subharmonic. An USC function u in Ω is called $\mathcal{G}$-subharmonic if for every C^2 function v which is $\widetilde{\mathcal{G}}$-subharmonic,

$$u + v \quad \text{is subaffine.}$$

Here are some properties of $\mathcal{G}$-subharmonic functions in Ω. The second is crucial in the use of the Perron method.

1. If u is $\mathcal{G}$-subharmonic and $D^2u(x)$ exists at some x then $D^2u(x) \in \mathcal{G}$.
2. If u, v, are $\mathcal{G}$-subharmonic then $\max(u; v)$ is $\mathcal{G}$-subharmonic.
3. A uniform limit of a sequence of functions which are $\mathcal{G}$-subharmonic; converging uniformly on compact subsets, is $\mathcal{G}$-subharmonic.
4. If H is a family of $\mathcal{G}$-subharmonic functions, locally uniformly bounded above, then the upper envelope.

$$u = \sup_{v \in H} \quad \text{has a USC regularisation } u^*$$

$$\text{and } u^* \text{ is } \mathcal{G}\text{-subharmonic.}$$

5. If u_1, u_2 are respectively $\mathcal{G}_1$-subharmonic, $\mathcal{G}_2$-subharmonic then

$$u_1 + u_2 \quad \text{is} \quad (\mathcal{G}_1 + \mathcal{G}_2)\text{-subharmonic}.$$

With this notation we now come to the definition of a solution of the problem, a viscosity solution:

Definition 2. As USC function u is a solution if u is $\mathcal{G}$-subharmonic and $-u$ is $\widetilde{\mathcal{G}}$-subharmonic (this means u is $\mathcal{G}$-superharmonic).

Next we take up the corresponding notion of "convexity" of Ω. First, to Ω is associated a cone $\overrightarrow{\mathcal{G}}$ obtained as follows: Consider any symmetric matrix B outside $\mathcal{G}$ and take all the rays from B that eventually enter $\mathcal{G}$. Their closure is a cone. Move it to have vertex at the origin; we obtain $\overrightarrow{\mathcal{G}}$; it is independent of the choice of B. $\overrightarrow{\mathcal{G}}$ is called the ray set of $\mathcal{G}$.

We consider bounded Ω with smooth boundary

Definition 3. Ω is called strictly $\overrightarrow{\mathcal{G}}$-convex if near every boundary point of Ω there is a local C^2 defining function ρ:

$$\rho = 0, \quad \nabla\rho \neq 0 \quad \text{on} \quad \partial\Omega$$
$$\rho < 0 \quad \text{in} \quad \Omega,$$

such that $D^2\rho \in int.\overrightarrow{\mathcal{G}}$.

Proposition 4. *If Ω is strictly $\mathcal{G}$-convex then there exists a global defining function ρ for $\partial\Omega$ which is strictly $\overrightarrow{\mathcal{G}}$-subharmonic on $\overline{\Omega}$. Furthermore $\exists\, C_0 > 0$ such that*

$$C\rho \quad \text{is } \mathcal{G}\text{-subharmonic for all} \quad C \geq C_0 .$$

Now, the existence theorem for given domain Ω and Dirichlet set $\mathcal{G}$.

Theorem 5. *Suppose $\partial\Omega$ is both $\overrightarrow{\mathcal{G}}$ and $\overrightarrow{\widetilde{\mathcal{G}}}$ strictly convex. Then for every $\phi \in C(\partial\Omega)$, there exists $u \in C(\overline{\Omega})$ which is a solution of the problem in Ω with $u = \phi$ on $\partial\Omega$. Furthermore, u is unique.*

As I mentioned, existence of the solution u is proved by the Perron method. Namely, u is the sup of all USC functions v which are $\mathcal{G}$-subharmonic, and $v \leq \phi$ on $\partial\Omega$.

To prove that this u is a solution, one proves that it is also a $\mathcal{G}$-super-solution, *i.e.* it is $\widetilde{\mathcal{G}}$-subharmonic.

A crucial step to do this is the

Subaffine Theorem. *If u and v are, respectively $\mathcal{G}$-subharmonic and $\widetilde{\mathcal{G}}$-subharmonic USC functions, then $u + v$ is subaffine.*

The proofs rely on max. princ. and make use of quasi-convex approximations u_ϵ (*i.e.*, $u_\epsilon + C|x|^2$ is convex). The condition that Ω is both $\overrightarrow{\mathcal{G}}$ and $\overrightarrow{\widetilde{\mathcal{G}}}$ strictly convex is used in constructing suitable barrier functions.

Incidentally those who learned the Perron method to solve the Dirichlet problem for Laplace equation may recall that at some point one used the fact that one could solve it in a ball. Such a step does not enter in [6]. Indeed, in the case of the Laplace equation, [6] proves the existence of a continuous solution. To prove, further, that it is smooth in Ω, one can use the fact that one can solve the equation in a ball for a smooth harmonic function with given continuous boundary values. Then, by the uniqueness of the [6] solution in that ball, it follows that the viscosity solution is smooth in the domain.

Here are a few examples of nonlinear equations that they solve

1. The homogeneous Monge-Ampère equation

$$\prod \lambda_i = \det(u_{ij}) \equiv 0 \quad \text{in} \quad \Omega$$
$$u = \phi \quad \text{on} \quad \partial\Omega .$$

In [2] the problem was treated in a convex domain Ω with $\partial\Omega$ in $C^{1,1}$, and with $\phi \in C^{1,1}$. It was proved that there is a $C^{1,1}$ convex solution *i.e.* a function satisfying

$$\lambda_1(D^2u(x)) = 0 \quad \forall \quad x \text{ in } \Omega.$$

In [6] they obtain n solutions, namely, for each j, they solve the equation

$$\lambda_j = 0. \tag{7}$$

This does not contradict the assertion that the solution is unique, for uniqueness is relative to the given $\mathcal{G}$. The $\mathcal{G}$ for the problem (7) is the set of symmetric matrics with $\lambda_j \geq 0$.

2. They solve the Dirichlet problem for

$$\lambda_j + \ldots + \lambda_q = 0$$

in a corresponding "convex" domain.

3. They solve the Dirichlet problem for a function u satisfying

$$u_{11} \cdot u_{22} = 0 \quad \text{and} \quad u_{11} \geq 0, \quad u_{22} \geq 0.$$

Here, strict "convexity" of Ω is a bit stronger than the statement that $\mathcal{G}$ is convex in the x_1 and x_2 directions. The unique solution is determined in each slice: $(x_3, \ldots, x_n)$ are constant. One cannot expect the solution to be smooth.

I include two results about convex functions that are used in the paper.

Proposition 6 (A. D. Alexandroff). *A convex function has second derivatives almost everywhere.*

The second result concerns Z. Slodkowski's notion of the largest eigenvalue of the Hessian of a convex function v:

Definition 7. Suppose v is differentiable at x, set

$$K(v, x) = \overline{\lim_{\epsilon \to 0}} \frac{2}{\epsilon^2} \sup_{|y|=1} [v(x + \epsilon y) - v(x) - \epsilon \nabla v(x) \cdot y].$$

(This plays the role of $\lambda_n(D^2v)$). If v is not differentiable at x, define

$$K(v, x) = +\infty.$$

Theorem 8 (Slodkowski [7]). *If v is convex in an open set and*

$$K(v, x) \geq C \quad a.e.$$

then the inequality holds everywhere.

[6] is a beautiful paper. It contains applications in geometric problems and complex manifolds.

It would be worth investigating under what additional conditions one can obtain smooth solutions.

References

[1] L. CAFFARELLI, L. NIRENBERG and J. SPRUCK, *The Dirichlet problem for nonlinear second order elliptic equations. III Function of the eigenvalues of the Hessian*, Acta Math. **155** (1985), 261–301.

[2] L. CAFFARELLI, L. NIRENBERG and J. SPRUCK, *The Dirichlet problem for the degenerate Monge-Ampère equation*, Revista Mat. Iberoamericana **2** (1980), 19–27.

[3] B. GUAN and J. SPRUCK, *Boundary value problems on S^n for surfaces of constant Gauss curvature*, Ann. of Math. **138** (1993), 601–624.

[4] B. GUAN, *The Dirichlet problem for Monge-Ampère equations in nonconvex domains and spacelike hypersurfaces of constant scalar curvature*, Trans. AMS **350** (1998), 4955–4971.

[5] B. GUAN, *The Dirichlet problem for Hessian equations on Riemannian manifolds*, Calc. of Var. and PDE's **8** (1999), 45–69.

[6] F. REESE HARVEY and H. BLAINE LAWSON JR., *Dirichlet duality and the nonlinear Dirichlet problem*, Comm. Pure Appl. Math. **62** (2009), 396–443.

[7] Z. SLODKOWSKI, *The Bremermann-Dirichlet problem for q-plurisubharmonic functions*, Ann. Scuola Norm. Sup. Pisa Cl. Sc. (4) **11** (1984), 303–326.

Fotocomposizione "CompoMat" Loc. Braccone, 02040 Configni (RI) Italy
Finito di stampare nel mese di maggio 2009
dalla CSR, Via di Pietralata, Roma

COLLOQUIA

The volumes of this series reflect lectures held at the "Colloquio De Giorgi" which regularly takes place at the Scuola Normale Superiore in Pisa. The Colloquia address a general mathematical audience, particularly attracting advanced undergraduate and graduate students.

Published volumes

1. Colloquium De Giorgi 2006. ISBN 978-88-7642-212-6
2. Colloquium De Giorgi 2007 and 2008. ISBN 88-7642-344-4